The Farm Process Version 2 (FMP2) for MODFLOW-2005—Modifications and Upgrades to FMP1

By Wolfgang Schmid, Department of Hydrology and Water Resources, University of Arizona, and

R.T. Hanson, U.S. Geological Survey

Office of Groundwater, Groundwater Resources Program

Techniques and Methods 6-A-32

U.S. Department of the Interior
U.S. Geological Survey

U.S. Department of the Interior
KEN SALAZAR, Secretary

U.S. Geological Survey
Suzette M. Kimball, Acting Director

U.S. Geological Survey, Reston, Virginia: 2009

For more information on the USGS—the Federal source for science about the Earth, its natural and living resources, natural hazards, and the environment, visit http://www.usgs.gov or call 1-888-ASK-USGS

For an overview of USGS information products, including maps, imagery, and publications, visit http://www.usgs.gov/pubprod

To order this and other USGS information products, visit http://store.usgs.gov

Suggested citation:
Schmid, Wolfgang, and Hanson, R.T., 2009, The Farm Process Version 2 (FMP2) for MODFLOW-2005—Modifications and Upgrades to FMP1: U.S. Geological Survey Techniques and Methods 6-A-32, 102 p.

Preface

This report presents a computer program for simulating the consumption and redistribution of surface water and groundwater from farming in the U.S. Geological Survey (USGS) hydrologic model, MODFLOW. The performance of this computer program has been tested in models of hypothetical groundwater flow systems; however, future applications of the programs could reveal errors that were not detected in the test simulations. Users are requested to notify the USGS if errors are found in the report or in the computer program. Correspondence regarding the report or program should be sent to:

U.S. Geological Survey
Water Resources Discipline
4165 Spruance Road, Suite 200
San Diego, CA 92101
Attention: Randall T. Hanson

Although this program has been used by the USGS, no warranty, expressed or implied, is made by the authors, the USGS, or the United States Government as to the accuracy and functioning of the program and related program material. Nor shall the fact of distribution constitute any such warranty, and no responsibility is assumed by the authors or the USGS in connection there with.

The computer program described herein consists in part of copyrighted scientific methodologies obtained from the copyright holder (Schmid, 2004). The copyright holder has granted full quotation, copy and use of these methods to the USGS and to the public. Requests for modification of copyrighted methods and for publication of such can be made to the copyright holder or to the address listed above.

The computer program documented in this report is part of the MODFLOW-2005 groundwater flow model. These and other groundwater programs are available from the USGS at the World Wide Web address:

http://water.usgs.gov/software/
or:
http://water.usgs.gov/software/lists/groundwater/

Contents

Contents—Continued

Figures

Figures—Continued

Tables

Conversions, Abbreviations, Acronyms, and Variable Definitions

SI to Inch/Pound

Multiply	By	To obtain
	Length	
centimeter (cm)	0.3937	inch (in.)
meter (m)	3.281	foot (ft)
	Flow rate	
cubic meter per day (m³/day)	0.00081071	acre-foot per day (acre-ft/d)

Temperature in degrees Celsius (°C) may be converted to degrees Fahrenheit (°F) as follows:

$$°F=(1.8×°C)+32$$

Temperature in degrees Fahrenheit (°F) may be converted to degrees Celsius (°C) as follows:

$$°C=(°F-32)/1.8$$

Elevation as used in this report refers to distance above the vertical datum.

*Transmissivity: The standard unit for transmissivity is cubic foot per day per square foot times foot of aquifer thickness [(ft³/d)/ft²]ft. In this report, the mathematically reduced form, foot squared per day (ft²/d), is used for convenience.

NOTE TO USGS USERS: Use of hectare (ha) as an alternative name for square hectometer (hm²) is restricted to the measurement of small land or water areas. Use of liter (L) as a special name for cubic decimeter (dm³) is restricted to the measurement of liquids and gases. No prefix other than milli should be used with liter. Metric ton (t) as a name for megagram (Mg) should be restricted to commercial usage, and no prefixes should be used with it.

Abbreviations, Acronyms, and Variable Definitions

Abbreviations, acronyms, and variables are defined as follows unless already defined in the FMP1 user guide (Schmid and others, 2006) and in section 'Data Input Instructions for FMP1 and New FMP2 Features:'

ASCII	American Standard Code for Information Interchange
CIMIS	California Irrigation Management Information System
$\alpha(\psi)$	dimensionless water uptake stress response function [-] ($0 \leq \alpha \leq 1$)
ψ	pressure head [L]
AURZ	Active unsaturated root zone [L]
ASRZ1	Active saturated root zone with maximum uptake [L]
ASRZ2	Active saturated root zone with reduced uptake [L]
ETR or ET_{ref}	Reference evapotranspiration flux [L/T]
FINF	Applied infiltration rate [L/T]
f_w	wetted fraction
HYDMOD	Computer program for calculating hydrograph time series data for MODFLOW
IUZFBND	Integer array defining active model area in which recharge and discharge are simulated.
K_c	Crop coefficient [-]
MF2005	MODFLOW-2005
MF2005-FMP2	MODFLOW-2005 version 1.6 with the Farm Process version 2
MNW	Multi-Node Well Package
MULT	Multiplier Package
PET	Potential evapotranspiration in the UZF1 Package
SFR	Streamflow Routing Package (SFR2 refers to the most recent published version 2; note: current code GWF1SFR7.F refers to code version 7)
SUB	Subsidence Package
UZF	Unsaturated Zone Flow Package
ZONEBUDGET	Computer program for calculating subregional water budgets for MODFLOW

The Farm Process Version 2 (FMP2) for MODFLOW-2005— Modifications and Upgrades to FMP1

By Wolfgang Schmid[1] and R.T. Hanson[2]

Abstract

The ability to dynamically simulate the integrated supply-and-demand components of irrigated agricultural is needed to thoroughly understand the interrelation between surface water and groundwater flow in areas where the water-use by vegetation is an important component of the water budget. To meet this need, the computer program Farm Process (FMP1) was updated and refined for use with the U.S. Geological Survey's MODFLOW-2005 groundwater-flow model, and is referred to as MF2005-FMP2. The updated program allows the simulation, analysis, and management of nearly all components of human and natural water use. MF2005-FMP2 represents a complete hydrologic model that fully links the movement and use of groundwater, surface water, and imported water for water consumption of irrigated agriculture, but also of urban use, and of natural vegetation. Supply and demand components of water use are analyzed under demand-driven and supply-constrained conditions. From large- to small-scale settings, the MF2005-FMP2 has the unique set of capabilities to simulate and analyze historical, present, and future conditions. MF2005-FMP2 facilitates the analysis of agricultural water use where little data is available for pumpage, land use, or agricultural information. The features presented in this new version of FMP2 along with the linkages to the Streamflow Routing (SFR), Multi-Node Well (MNW), and Unsaturated Zone Flow (UZF) Packages prevents mass loss to an open system and helps to account for "all of the water everywhere and all of the time."

The first version, FMP1 for MODFLOW-2000, is limited to (a) transpiration uptake from unsaturated root zones, (b) on-farm efficiency defined solely by farm and not by crop type, (c) a simulation of water use and returnflows related only to irrigated agriculture and not also to non-irrigated vegetation, (d) a definition of consumptive use as potential crop evapotranspiration, (e) percolation being instantly recharged to the uppermost active aquifer, (f) automatic routing of returnflow from runoff either to reaches of tributary stream segments adjacent to a farm or to one reach nearest to the farm's lowest elevation, (g) farm-well pumping from cell locations regardless of whether an irrigation requirement from these cells exists or not, and (h) specified non-routed water transfers from an undefined source outside the model domain.

All of these limitations are overcome in MF2005-FMP2. The new features include (a) simulation of transpiration uptake from variably saturated, fully saturated, or ponded root zones (for example, for crops like rice or riparian vegetation), (b) definition of on-farm efficiency not only by farm but also by crop, (c) simulation of water use and returnflow from non-irrigated vegetation (for example, rain-fed agriculture or native vegetation), (d) use of crop coefficients and reference evapotranspiration, (e) simulation of the delay between percolation from farms through the unsaturated zone and recharge into the uppermost active aquifer by linking FMP2 to the UZF Package, (f) an option to manually control the routing of returnflow from farm runoff to streams, (g) an option to limit pumping to wells located only in cells where an irrigation requirement exists, and (h) simulation of water transfers to farms from a series of well fields (for example, recovery well field of an aquifer-storage-and-recovery system, ASR).

In addition to the output of an economic budget for each farm between irrigation demand and supply ("Farm Demand and Supply Budget" in FMP1), a new output option called "Farm Budget" was created for FMP2, which allows the user to track all physical flows into and out of a water accounting unit at all times. Such a unit can represent individual farms, farming districts, natural areas, or urban areas.

The example model demonstrates the application of MF2005-FMP2 with delayed recharge through an unsaturated zone, rejected infiltration in a riparian area, changes in demand owing to deficiency in supply, and changes in multi-aquifer pumpage owing to constraints imposed through the Farm Process and the MNW Package.

[1]Research Hydrologist, Department of Hydrology and Water Resources, University of Arizona.

[2]U.S. Geological Survey

Introduction

The estimation of irrigation demand from a combination of surface water and groundwater is needed to accurately simulate groundwater flow in areas where the water-use by vegetation is an important component of the water budget. To meet this need, the computer program Farm Process (FMP) (Schmid, 2004; Schmid and others, 2006) was developed for use with the U.S. Geological Survey's finite-difference groundwater-flow model, commonly called MODFLOW. The first version of the Farm Process (FMP1) was developed for use with MODFLOW-2000 (Harbaugh and others, 2000). Because MODFLOW-2000 (Harbaugh and others, 2000) was updated to MODFLOW-2005 (Harbaugh, 2005), FMP1 had to be updated to facilitate the use of the program with the new version of MODFLOW.

There also was a need to extend the capabilities of analyzing the use and movement of water resources throughout the hydrologic cycle in a process-based context within applications of MODFLOW. This requires including the simulation of fully coupled use and movement of water on the land surface within a fully coupled hydrologic model such as MODFLOW-2005 (Harbaugh, 2005). The simulation of all water includes the application, consumption, and movement of water for natural vegetation, agriculture, and urban settings on the land surface (the uppermost surface of the hydrologic model hereinafter referred to as the "landscape"). The use and movement of inflows and outflows derived from precipitation, surface water and groundwater are now facilitated within a fully coupled, process-based hydrologic simulation model through the integration of the Farm Process version 2 within MODFLOW-2005 version 1.6, hereinafter referred to as MF2005-FMP2. MF2005-FMP2 represents both an integration of FMP into the newest version of MODFLOW and an updated and upgraded version of the Farm Process (FMP2). FMP2 is linked to the Streamflow Routing (SFR2; Niswonger and Prudic, 2005), Multi-Node Well (MNW; Halford and Hanson, 2002), and Unsaturated Zone Flow (UZF1; Niswonger and others, 2006) Packages, which helps to account for all water in the simulated system.

FMP1 provided a full coupling of the fundamental components of irrigation such as plant consumptive use, evaporation, effective precipitation, surface-water delivery, supplemental pumpage of groundwater, irrigation runoff, and deep percolation of excess applied irrigation through the partially saturated root zone. FMP1 facilitated the simulation and analysis of conjunctive use of surface water and groundwater which is commonly complex for most modern agricultural settings. Pumpage, required to simulate groundwater flow in agricultural areas, is not always metered, which complicates the simulation of conjunctive use. Integrated estimation of transpiratory and evaporative consumptive use of precipitation as well as applied surface-water and groundwater deliveries is needed to estimate a complete water-use budget and related inflow and outflow components that are connected to the surface-water and groundwater flow systems for both anthropogenic (that is, agricultural and urban) and natural settings. In addition, it is critical that the water-use budget be implicitly linked to these flow systems through interdependencies of flow-dependent and head-dependent relations that allow the consumptive requirements of the crops and the natural evaporation from the soils to be dependent on the water level in the uppermost aquifer. For example, in FMP1 the heads in aquifers below the irrigated agriculture are coupled to the evaporation and transpiration that occur in the root zone of the soils on a cell-by-cell basis. The simulation of inflows to and outflows from water-balance accounting units (hereinafter referred to as "farms" or, for aggregations of multiple farms or irrigation districts, "virtual farms") and groundwater systems also should represent fallow or noncultivated regions, as well as urban irrigation located within the modeling domain.

The new features of FMP2 facilitate the extended simulation and analysis of conjunctive use and helps the user track the use and movement of water throughout the modeling domain and simulation period or, in common terms, 'all the water everywhere and all of the time' (Hanson and others, 2008a).. These new capabilities include (a) simulation of transpiration uptake from variably saturated, fully saturated, or ponded root zones (for example, for crops like rice or riparian vegetation), (b) definition of on-farm efficiency not only by farm but also by crop, (c) simulation of water use and returnflow from non-irrigated vegetation (for example, rain-fed agriculture or native vegetation), (d) use of crop coefficients and reference evapotranspiration, (e) simulation of the delay between percolation from farms through the unsaturated zone and recharge into the uppermost active aquifer by linking FMP2 to the UZF1Package, (f) an option to manually control the routing of returnflow from farm runoff to streams, (g) an option to limit pumping to wells located only in cells where an irrigation requirement exists, and (h) simulation of water transfers to farms from a series of well fields (for example, recovery well field of an aquifer-storage-and-recovery system, ASR).

Regional hydrologic models frequently are used to analyze the complex relations between simulated and measured supply and demand components of conjunctive use systems. This requires simulation of the interdependency of head-dependent flows (fig. 1 and 2). Several Packages and processes of MODFLOW not only are linked to groundwater flow but also are linked to one other. This linkage causes flow terms of each respective Package or process to depend mutually on flow terms. The structure of MODFLOW-2005 has made it significantly easier to use several linked MODLFOW Packages and processes in the simulation of these interdependent flow components and related constraints on the supply and demand components of conjunctive use systems in FMP2 (fig. 1 and fig. 2). The mutual effect that the linked flow terms have on one another in the context of conjunctive use can now be assessed through surface landscape water-accounting units (for example, individual farms or water balance regions in FMP2) or subsurface groundwater accounting units (by ZONEBUDGET, Harbaugh, 1990). The groundwater accounting units may be linked to the surface accounting units that overlay them. The networks of interdependencies that are facilitated in the new MF2005-FMP2 are illustrated in figure 1 in the context of constrained deliveries and returnflows within the hydrologic cycle of a typical flow system. These interdependencies include the supply-derived constraints of interdependent streamflows and surface-water deliveries between FMP2 and SFR2 (Niswonger and Prudic, 2005). Additionally, the pumping capacity (hydraulic limit) of FMP2, combined with optional user-specified constraints from MNW (head and drawdown limits; Halford and Hanson, 2002), provide supply-derived constraints from the groundwater deliveries. Similarly, there are constraints within the movement of water on inefficient losses to runoff returnflow and percolation. Constraints on the percolation of infiltration from farm root zones into the deeper unsaturated zone can optionally be imposed through a linkage between FMP2 and the UZF1 Package (Niswonger and others, 2006). The constraints on returnflows may indirectly impact groundwater or surface-water supply by delaying groundwater recharge or increasing streamflow through rejected infiltration back to the SFR2 Package stream network (fig.1). Surface-water supply may further be constrained by the location of runoff returnflow back to the SFR2 Package stream network.

The new linkage between FMP2 and UZF1 that is added to the existing linkages between FMP2 and SFR2 or MNW is an example of a component of the FMP2 framework that facilitates the integrated simulation of "all the water everywhere and all of the time." In this relation, deliveries to farms may exceed the actual demand for crop irrigation and represent diversions that historically were used, in part, to "perfect" surface-water rights. Inefficient losses that result from excess demands for water supply can be returned to the groundwater or streamflow-routing system to become a potential source of supply for other users. If all flows and uses of water are not tracked throughout the hydrologic cycle all of the time, the simulation can sustain considerable additional uncertainty or inaccuracies through the loss of water within the simulation framework.

Purpose and Scope

The purpose of this report is to describe the new capabilities or features of FMP2 and associated concepts and general data requirements for new FMP2 features. The data input instructions for the existing unchanged FMP1 input items and the modified or new FMP2 input items also are included. A hypothetical example problem is used to illustrate some of the features of FMP2 needed for many regional hydrologic models that include simulations of supply and demand related to irrigated agriculture, natural vegetation and urban setting of water use and movement. The example problem is simulated with MF2005 using FMP2 jointly with the SFR2, UZF1, and MNW Packages to demonstrate the new linkages and flow interdependencies now available in MF2005-FMP2. The example demonstrates the application of FMP2 with delayed recharge through an unsaturated zone utilizing the UZF1 Package, rejected infiltration in a riparian area, changes in demand owing to deficiency in supply, and changes in multi-aquifer pumpage owing to constraints imposed through FMP2 and the MNW Package.

Many of the new FMP2 features and additional modifications of other selected packages were required to align the functionality of MODFLOW-2000 and FMP1 with the hydrologic and geologic architecture of recent applications to regional hydrologic systems such as the Central Valley, California (Faunt and others, 2008a,b, 2009a,b,c) and the Pajaro Valley, California (Hanson and others, 2008b). The additional packages that were modified or updated for this purpose include the Multiplier (MULT) and the Hydrograph Time Series (HYDMOD) (Schmid and Hanson, 2009) Packages. MF2005-FMP2 includes but is not limited to all the new or modified features of FMP1, MULT, and HYDMOD that were made for the applications mentioned above. In addition, the SFR2, UZF1, and SUB packages of MF2005 were slightly modified. All modifications of MULT, HYDMOD, SFR2, UZF1, and SUB are summarized in Appendix B.

Acknowledgments

The development of FMP2 was supported by the USGS Office of Groundwater and USGS the California Water Science Center. The authors would like to acknowledge Stan Leake for the contribution of revisions to the subsidence package (Appendix B). The authors would like to thank the reviewers Phil King, Steve Peterson, Devin Galloway, and Peter Martin, as well as Larry Schneider for the illustrations.

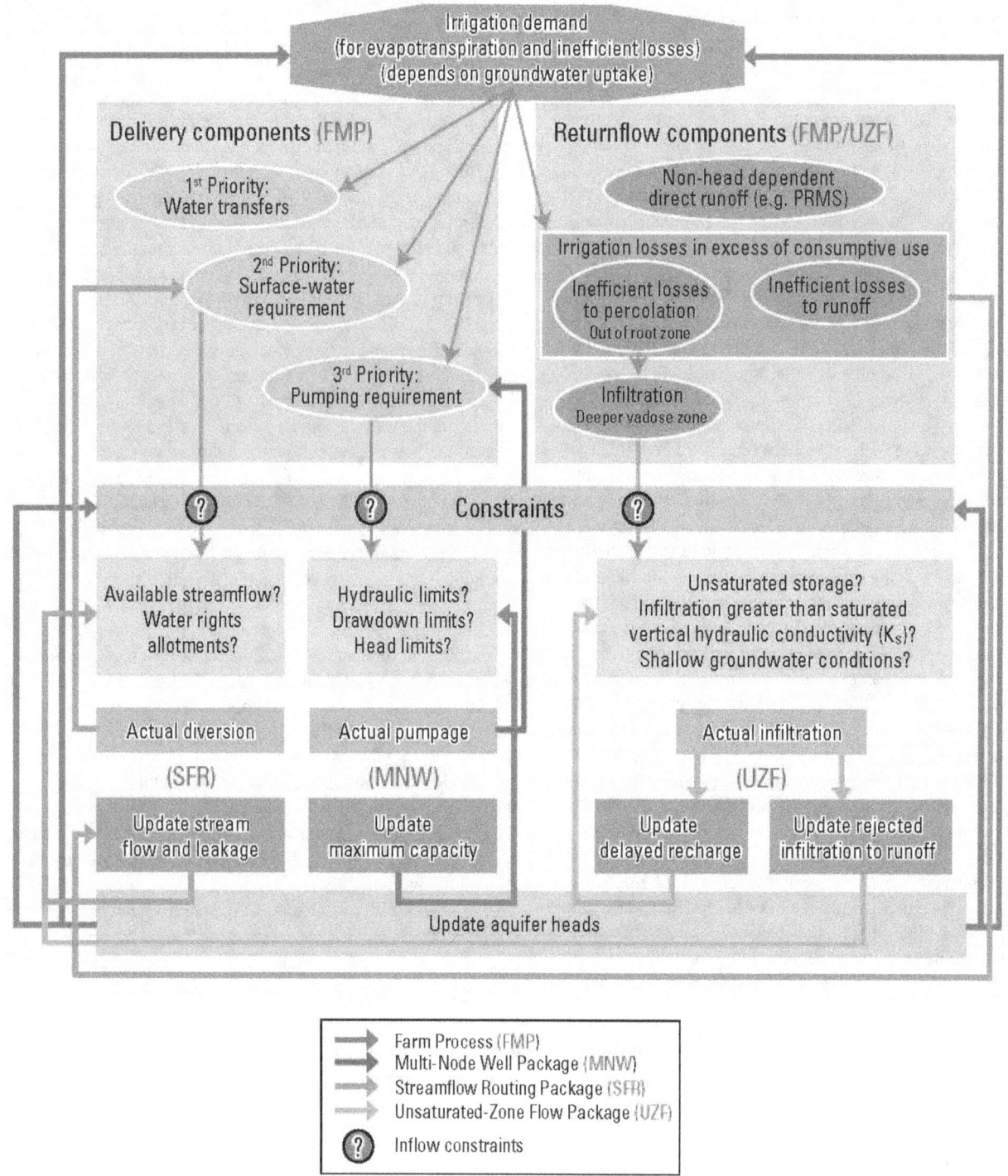

Figure 1.   Interdependencies within MF2005-FMP2 and the related constraints on the supply and demand components.

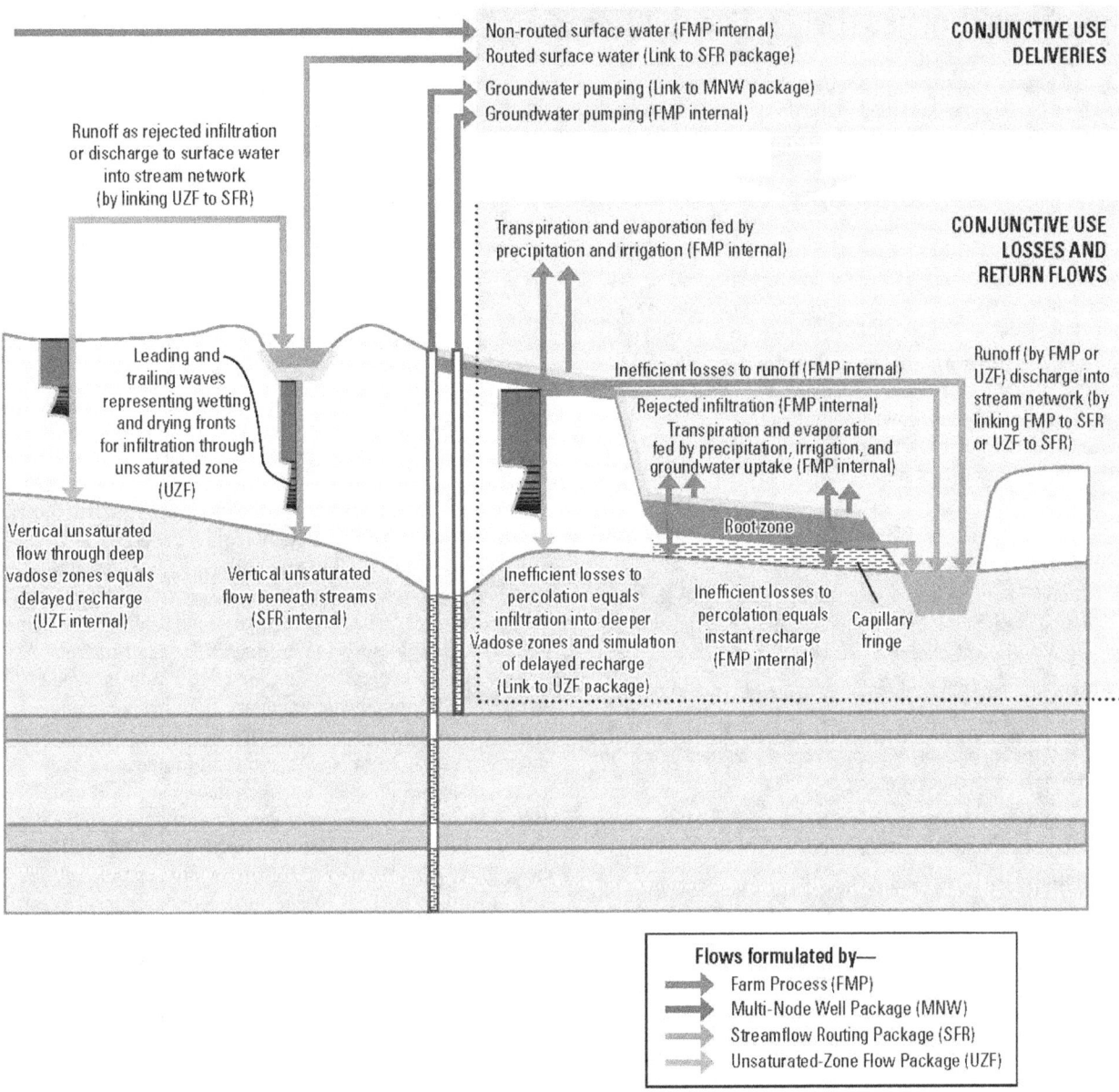

Figure 2. Interdependent flows within a hydrologic system simulated by MF2005-FMP2.

New Farm Process (FMP2) Capabilities

Additional features of and improvements made to the Farm Process are described in the order, in which they affect the FMP input data set. Although most of the new or amended features simply require a change in user-specified options (hereinafter referred to as "flag settings" that control input options through user specification in the FMP input file) in Item 2 of the FMP input data set, some features do not require changes to the flags but involve direct changes to the data sets that are read for the entire simulation or for every stress period. Hence, the order in which the new and amended features are described in this report is not related to the level of significance of each feature. For instance, the new analytical concept that allows the simulation of root uptake under variably saturated conditions is probably the most extensive and significant improvement of MF2005-FMP2, but is described last.

The additional features and improvements made to FMP2 include:

1. Simulation of additional consumptive use types that include estimation of consumptive use as the product of crop coefficients and reference evapotranspiration;

2. Simulation of delayed recharge and additional runoff components through linkage of FMP2 to UZF1;

3. Fully-routed deliveries from diversion segments or from any type of stream segment;

4. Additional coordinate options for the specification of a point of diversion for semi-routed deliveries;

5. Fully-routed runoff returnflows from non-diversion segments or from any type of segment;

6. Specification of semi-routed returnflows to specific reaches of the stream network;

7. Output of Farm Budget;

8. Output of information about the routing of farm deliveries and farm returnflows;

9. Restrictions on farm-well pumping, where no irrigation requirement exists;

10. Farm irrigation supply through series of priority-ranked well fields;

11. Recomputation of FMP2 flow terms that depend on flow terms of linked packages;

12. Closure criterion for MNW flow terms to converge to FMP pumping requirement;

13. Specification of on-farm efficiencies by farm and by crop as a matrix;

14. Simulation of use and movement of water on the landscape for non-irrigated vegetation;

15. Simulation of root uptake under variably saturated conditions.

With the capability to simulate root-zone pressure heads that span from negative to positive pressure heads, FMP2 can now simulate the growing of crops that take up water under saturated conditions such as rice or some types of riparian vegetation. A more distributed specification of on-farm irrigation efficiencies allows the user to specify efficiencies as a matrix by farm and by crop for the entire simulation or for each stress period. The addition of an input flag to the consumptive use data allows the differentiation between irrigated vegetation and non-irrigated vegetation. Non-irrigated vegetation can represent dry-land farming or natural vegetation settings such as rangeland, forests, or riparian settings. This flag prevents irrigation water from surface-water deliveries and groundwater pumpage from being applied to the non-irrigated vegetation, but still allows evaporation and transpiration from precipitation and groundwater and inefficient losses from excess precipitation to runoff and deep percolation. The expansion of consumptive-use features allows the specification of consumptive use as the product of the crop coefficients and reference evapotranspiration for each crop or a group of crops for the entire simulation or for each stress period. This allows the simulation of crop water demand during individual growth stages for each crop or crop group. Locations along the streamflow network, where runoff from farms can be returned to the streamflow network, can now be specified. The additional restriction of farm wells now allows pumpage to be restricted to farm wells that are located in cells, where vegetation requires irrigation water. The additional budget components facilitate output of a detailed time series of all the inflows to, and outflows from, each farm (water-balance subregion). The input data options and specifications for each of these new features are described in the following section.

Concepts and General Data Requirements for New FMP2 Features

The basic concepts and general data requirements of new FMP2 features that are add-on options to existing FMP1 features are described in the order of their occurrence within the input instructions in the Chapter "Data Input Requirements of FMP1 and new FMP2 Features." For new FMP2 features, the parameter and input item number of the FMP1 input instructions are referenced in parentheses after the respective section titles. The summary of input parameters (tables 1 and 2) includes the previous unchanged and changed or new input items and is listed in the Chapter "Data Input Requirements of FMP1 and new FMP2 Features."

Consumptive-Use Options

New consumptive use options require modifications to the Consumptive Use flag, ICUFL, specified in Item 2 of the FMP input data set and a new parameter ETR specified in a new item between previous Items 27 and 28 of the FMP input data set. In FMP1, the user had the option to provide the consumptive use for each crop as fluxes of the potential crop evapotranspiration, ETc-pot, for each stress period in Item 27. FMP2 lets the user now specify crop-specific crop coefficients, Kc, in Item 27 (now called 27a) and a spatially constant or distributed reference evapotranspiration in a new Item 27b (Faunt and others, 2009c). However, even if the user preferred to use the previous option of specifying potential crop evapotranspiration data, the reference evapotranspiration might still be a necessary data input for fallow cells not associated with a crop ID or crop specific consumptive use as read in Item 27a.

The use of previous and new options of the Consumptive Use flag, ICUFL, in Item 2 is explained as follows (tables 1 and 2):

ICUFL = 3: FMP2 calculates a daily potential crop evapotranspiration flux (ETc-pot) by multiplying a daily reference evapotranspiration flux (ETref) read as time series for the entire simulation in Item 16 with a daily crop coefficient Kc derived from parameters read for the entire simulation as Item 15 (ETc-pot = Kc × ETref). FMP2 multiplies a daily ETc-pot averaged over each time step by the area of each cropped cell (ICID(IC,IR) > 0) to yield a cell-by-cell ETc-pot flow rate for each time step. FMP2 multiplies the daily ETref flux averaged over each time step by the area of each fallow cell (ICID(IC,IR) = –1) to yield a cell-by-cell ETref flow rate for each time step. The ETref is assumed to be 100 percent evaporative for fallow cells where no transpiration exists.

ICUFL = 2: A list of crop specific fluxes of potential crop evapotranspiration (ETc-pot) is read as Item 27a (Crop-ID, ETc-pot flux) for every stress period. FMP2 multiplies this ETc-pot flux by the area of the each cropped cell (ICID(IC,IR) > 0) to yield a cell-by-cell ETc-pot flow rate for each stress period. FMP2's fallow-cell option (ICID(IC,IR) = –1) cannot be used because no ETref flux is read if ICUFL = 2.

ICUFL = 1: (new) A list of crop specific fluxes of potential crop evapotranspiration (ETc-pot) is read as Item 27a (Crop-ID, ETc-pot flux) for every stress period and a constant or 2D real array reference evapotranspiration ETref (NCOL,NROW) is read as Item 27b for every stress period. FMP2 multiplies the ETc-pot flux by the area of the cropped cell (ICID(IC,IR) > 0) to yield a cell-by-cell ETc-pot flow rate for each stress period. FMP2 multiplies the ETref flux by the area of each fallow cell (ICID(IC,IR) = –1) to yield a cell-by-cell ETref flow rate for each stress period. The ETref is assumed to be 100 percent evaporative for fallow cells where no transpiration exists.

ICUFL = –1: (new) A list of crop specific crop coefficients (Kc) is read as Item 27a (Crop-ID, Kc) for every stress period and a constant or 2D real array of reference evapotranspiration ETref (NCOL,NROW) is read as Item 27b for every stress period. FMP2 multiplies the Kc by the ETref flux and by the area of each cropped cell (ICID(IC,IR) > 0) to yield a cell-by-cell ETc-pot flow rate for each stress period. FMP2 multiplies the ETref flux by the area of each fallow cell (ICID(IC,IR) = –1) to yield a cell-by-cell ETref flow rate for each stress period. The ETref is assumed to be 100 percent evaporative for fallow cells where no transpiration exists.

A limitation to the previous concept of using ETc-pot fluxes and to the new concept of using Kc values is that data must be derived from non-stressed conditions (for example, Allen and others, 1998). If stressed conditions do exist when gathering or deriving ETc-pot fluxes or Kc values, then the respective data need adjustment to reflect the altered conditions of root uptake and evapotranspiration that are due to moisture stress (Food and Agriculture Organization, 2007; Snyder and Eching, 2007). The user may use a scale factor as a water-stress coefficient to modify the ETc-pot fluxes (ICUFL = 2 or 1) or Kc values (ICUFL= -1). The scale factor must be preceded by the keyword "SFAC" and be entered in the line above the crop specific list of ETc-pot fluxes (ICUFL = 2 or 1) or Kc values (ICUFL = −1) for each stress period (Schmid and others, 2006, page 68). Alternatively, for a-priori stressed conditions, the user may use the simpler crop consumptive use Concept 2 (ICCFL = 2), which does not account for plant- and soil-specific anoxia, and read stressed evapotranspiration as "consumptive use," CU, in Item 27a. However, even though the input consumptive use data may account for some stressed conditions (for example, through anoxic conditions for certain ranges of negative pressure heads in the unsaturated root zone), FMP2 will reduce root uptake for most crops under saturated conditions as a result of a rising water level. Actual evapotranspiration input data that account for stresses to water uptake under both unsaturated and saturated conditions would falsely assume that the model evapotranspiration is independent of the water-level elevation. A better way of using available actual evapotranspiration data derived under stressed conditions is to calibrate the model's simulated actual evapotranspiration against those data. The ratio between a calibrated actual crop evapotranspiration and the potential crop evapotranspiration allows the derivation of a time-varying stress factor.

Delayed Recharge and Additional Runoff Components by Linkage to UZF Package

A new link between FMP2 and the Unsaturated-Zone Flow (UZF1) Package (Niswonger and others, 2006) required an extension of the Consumptive-Use Concept flag, ICCFL, specified in Item 2 of the FMP input data set. In FMP2, percolation is simulated as inefficient losses of irrigation or precipitation in excess of its respective consumptive use. The previous approach used in FMP1 assumes no delay between percolation beyond the bottom of the root zone and recharge to the uppermost active aquifer. A new linkage between FMP2 and the Unsaturated-Zone Flow (UZF1) Package allows the use of percolation as applied infiltration below farms (or water accounting units). The linkage is activated by switching the ICCFL flag in FMP2 to 3 when using Consumptive-Use Concept 1 and to 4 when using Consumptive-Use Concept 2 (Schmid and others, 2006, pages 11–14 and 70).

For farms simulated by FMP2, transpiration and evaporation are simulated by FMP2 as separate components of consumptive use prior to calculating the inefficient losses to percolation. Therefore, when linking FMP2 and UZF1, the UZF1 evapotranspiration flag, IETFLG, has to be switched to zero to avoid a double accounting of evapotranspiration (Item 1 in Niswonger and others, 2006, pp. 29–30). If, in one model, some UZF1 infiltration areas are independent from FMP2 farms and some others coincide with FMP2 farms, then the IETFLG flag in UZF1 may be switched to 1 to allow the simulation of evapotranspiration for "stand-alone" UZF1 infiltration areas. However, the potential evapotranspiration PET in UZF1 has to be set to zero where FMP2 simulates transpiration and evaporation to avoid a double accounting of evapotranspiration (Item 12 in Niswonger and others, 2006, p. 32).

The FMP2-calculated percolation is passed on to and partitioned by the linked UZF1 Package into different components including various runoff components, actual infiltration into the deeper vadose zone below farms, unsaturated-zone storage below farms, and recharge. The UZF1 Package also accounts for groundwater discharge to land surface or streams and rejected infiltration caused by high groundwater levels. These overland runoff components are simulated in UZF1 infiltration areas that can be either independent from or coincide with FMP2 farms.

In the UZF1 Package, vertically downward flow through the unsaturated zone is simulated by a kinematic wave approximation to Richards' equation, which, in turn, is solved by the method of characteristics (Smith, 1983). The approach assumes that unsaturated flow occurs in response to gravity potential gradients only and ignores negative potential gradients; the approach further assumes uniform hydraulic properties in the unsaturated zone for each vertical column of model cells. The Brooks–Corey function is used to define the relation between unsaturated hydraulic conductivity and water content (Brooks and Corey, 1966).

Variables used by the UZF1 Package, defined in Niswonger and others (2006), include initial and saturated water contents, saturated vertical hydraulic conductivity, and an exponent in the Brooks-Corey function. Residual water content is calculated internally by the UZF1 Package on the basis of the difference between saturated water content and specific yield. To establish a link between FMP2 farms and UZF1 infiltration arrays, the ICCFL flag in FMP2 must be switched to 3 or 4, and, in addition, FMP2 farm identification arrays must coincide with UZF1 infiltration arrays. The areal extent of UZF1 infiltration areas is defined in UZF1 by the IUZFBND array (Item 2, Niswonger and others, p. 30). The UZF1 input instructions require the user to specify an applied infiltration rate, FINF (Item 10, Niswonger and others, p. 32), which, if FMP2 is linked to UZF1, may be specified as a 'place holder' number (for example, zero). The link passes on the FMP2-calculated percolation beyond the root zone to the UZF1 Package as quasi "applied infiltration" into the vadose zone below the root zone.

Like the specified applied infiltration rate for non-farm UZF1 infiltration areas, the percolation rate under FMP2 farm areas linked to UZF1 infiltration areas also may be limited by the saturated vertical hydraulic conductivity. The resulting excess infiltration, groundwater discharged to land surface, and rejected infiltration caused by high groundwater levels are three runoff components simulated by UZF1 and may be transmitted directly as inflow to specified streams or lakes if the SFR2 or Lake (LAK3) (Merritt and Konikow 2000) Packages are active. These UZF1 runoff components add to the FMP2-generated inefficient losses to surface-water runoff (figs. 1 and 2). The total runoff is summed for each farm and time step in the FMP2 output Farm Budget, FB_DETAILS.OUT, as rates in column "Q-run-out" or as cumulative volumes in column "V-run-out." In UZF1, for each cell's runoff, the user has to specify the number of a receiving stream segment or lake in an array called IRUNBND (Item 3 in Niswonger and others, 2006, p. 30); otherwise, this water is removed from the model.

This report includes an example in which MODFLOW-2005 with FMP2, UZF1, SFR2, and MNW was used to demonstrate the new linkages and flow interdependencies now available in MF2005-FMP2. The linkage to the UZF1 Package facilitates delayed recharge through the unsaturated zone and facilitates groundwater discharge to land surface in the riparian areas in the example problem described later.

The Consumptive-Use Concept Flag, ICCFL, in Item 2 of the FMP1 input instructions was expanded beyond the two options used for the approximation of evapotranspiration fluxes with two new options in FMP2 to indicate a link between FMP2 and the UZF1 Package as follows:

ICCFL = 1:   Plant-and soil-specific pseudo steady state transpiration approximated by analytical solution: A restriction of active root zone corresponding to anoxia- or wilting-related pressure heads is determined by FMP using analytical solutions of a vertical pseudo steady state pressure head distribution over the depth of the total root zone. (FMP2 not linked to UZF1).

ICCFL = 2:   Nonplant- and nonsoil-specific simplification of Concept 1 (FMP2 not linked to UZF1).

ICCFL = 3:   Plant-and soil-specific pseudo steady state transpiration approximated by analytical solution: A
(new)   restriction of active root zone corresponding to anoxia- or wilting-related pressure heads is determined by FMP using analytical solutions of a vertical pseudo steady state pressure head distribution over the depth of the total root zone. (FMP2 linked to UZF1: FMP2 farm identification arrays linked to coinciding UZF1 infiltration arrays).

ICCFL = 4:   Nonplant- and nonsoil-specific simplification of Concept 1. (FMP2 linked to UZF1: FMP2 farm
(new)   identification arrays linked to coinciding UZF1 infiltration arrays).

Fully-Routed Delivery

Fully-routed deliveries are defined in FMP as surface water that is routed directly to a farm by an open-channel conveyance network. Previously in FMP1, a farm could receive fully-routed surface water from the farthest upstream reach of a series of diversion-segment reaches adjacent to a farm boundary (if routed delivery flag IRDFL = 1). Such diversion segments of the SFR Package allow the simulation of canals or laterals that divert water from a main-stem river. In FMP1, the automatic generation of fully-routed deliveries from the farthest upstream reach of a diversion segment was generated only if that reach was adjacent to the boundary of a farm. "Adjacent" in this context meant that there had to be no model cell between at least one cell containing a diversion-segment reach and at least one cell of a particular farm.

Changes to the way fully-routed deliveries were conceptualized in FMP1 and a new option of fully-routed deliveries from any type of stream segments required modifications to the Fully-Routed Delivery flag, IRDFL, specified in Item 2 of the FMP input data set. In FMP2, the option to divert fully-routed deliveries from the uppermost reach of diversion segments "adjacent" to the farm (IRDFL = 1) was changed such that fully-routed deliveries from the uppermost reach of a series of diversion-segment reaches are possible only if those reaches are located "within" the cells that compose a particular farm. In addition, in FMP2, a new option was made available to allow farms to divert water from the uppermost reach of a series of reaches of any type of stream segment located within the farm (IRDFL = −1).

The use of modified and new options of the Fully-Routed Surface-Water Delivery flag, IRDFL, is explained as follows (tables 1 and 2):

IRDFL = 0:	No surface-water delivery exists.
IRDFL = 1: (modified)	Fully-routed surface-water delivery may occur from the uppermost reach of a series of diversion segment reaches located within a farm. Caution: Streamflow fully-routed through a conveyance network directly to a farm can occur only if (1) "SFR" is specified in Name File, and (2) at least one reach of a diversion segment is located within the farms, and (3) streamflow is available.
IRDFL = −1: (new)	Fully-routed surface-water delivery may occur from the uppermost reach of a series of reaches of any type of stream segment located within a farm. Caution: Streamflow fully-routed through a conveyance network directly to a farm can occur only if (1) "SFR" is specified in Name File, and (2) at least one reach of any type of segment is located within the farms, and (3) streamflow is available.

Clearly, the setup of the Farm ID array and the specification of stream reach locations (row, column, segment number, reach number) in the SFR Package decides whether fully-routed deliveries to farms can be simulated. For farms, where potential head-gate reaches are located outside the farm's domain, the user may specify remote points of diversion for farms for semi-routed deliveries (ISRDFL > 0). For farms, where potential head-gate reaches are located within the farm's domain, the user has the option to manually suppress an automatic fully-routed delivery (IRDFL = 0) and to manually specify points of diversion (ISRDFL > 0).

Semi-Routed Delivery

Semi-routed surface water is defined in FMP as surface water routed through an open-channel network to a point of diversion from where water then is delivered in a non-routed, closed-channel form (for example, by pipeline) to a farm that is remote from the point of diversion. Such a farm only can receive "semi-routed" but not "fully-routed" deliveries.

Changes to the way a point of diversion for a semi-routed delivery is specified required a modification to parameter REACH in Items 20 or 34 of the FMP input data set. Previously in FMP1, a stream-reach number required to specify a point of diversion for a semi-routed delivery was defined as a sequential number of a reach from zero to the total number of reaches that are active during the simulation as specified in the SFR1 input file (NSTRM, Item 1, Prudic and others, 2004, page 40). This option allowed defining the diversion point uniquely by just the reach number (option 5 in Schmid and others, pages 54, 61, and 78) but made it difficult for the user to keep track of the segment to which each reach belongs to. In FMP2, the reach number of the diversion point is now aligned with the reach numbering scheme per segment in the SFR2 input instructions (IREACH, Item 2, Niswonger and Prudic, 2005, page 20). A unique definition of a diversion point for complex cases, where multiple reaches may exist within one model cell, now requires the entry of both segment and reach numbers. Therefore, option 5 was deleted from FMP2.

Fully-Routed Runoff Returnflow

A new option of fully-routed runoff returnflow from either non-diversion segments or any type of stream segments located within a farm required a new Fully-Routed Runoff Returnflow flag, IRRFL, in Item 2 of the FMP input data set. Previously in FMP1, the surface-water runoff from a farm was automatically returned as fully-routed returnflow if a reach of a non-diversion segment was "adjacent" to the boundary of the farm. Analogous to the changes made to the fully-routed delivery option, in FMP2, fully-routed surface-water runoff is returned to reaches of a non-diversion segment only if they are located "within" the farm (IRRFL = 1). In addition, in FMP2, a new option was made available to allow the runoff of a farm to be prorated over reaches not only of a non-diversion segment but of any type of segment located within the farm (IRRFL = −1). As in FMP1, the user has to decide to which type of segment the runoff is returned if the reaches of the segment are located within the farm. If the SFR Package is specified in the Name File, IRRFL = 0 is not a valid entry because runoff would be computed in the physical farm budget, but would not be returned to the stream network. This would violate the assumption in FMP2 of accounting for 'all the water everywhere at all times.' If IRRFL = 0 is specified while the SFR Package also is specified in the Name File, FMP2 returns an error message to the listing file and stops the program.

The use of the new Fully-Routed Runoff Returnflow flag, IRRFL, is explained as follows (also, see tables 1 and 2):

IRRFL = 0:   No surface-water runoff returnflow possible (no "SFR" specified in Name File).

IRRFL = 1:   Surface-water runoff may be returned as fully-routed returnflow to a series of *non-diversion-segment reaches* located *within* a farm (prorated over each reach weighted by the reach length). Caution: Fully-routed returnflow directly from a farm to a series of non-diversion-segment reaches can occur only if (1) "SFR" is specified in Name File and (2) at least one reach of a non-diversion segment is located within the farm.

IRRFL = –1:   Surface-water runoff may be returned as fully-routed returnflow to a series of reaches of *any segment of stream type* located *within* a farm (prorated over each reach weighted by the reach length). Caution: Fully routed returnflow directly from a farm to a series of reaches of any type of stream segment can occur only if (1) "SFR" is specified in Name File and (2) at least one reach of a non-diversion segment is located within the farm.

Semi-Routed Runoff Returnflow

A new option of semi-routed runoff returnflow from any type of stream segments required a new Semi-Routed Returnflow flag, ISRRFL, in Item 2 of the FMP input data set and new parameters ROW COLUMN SEGMENT REACH specified as new Item 20b or 34b. ISRRFL allows the simulation of surface-water runoff returnflow from excess irrigation and/or excess precipitation that is allocated as non-routed returnflow to a point of recharge at a specified reach of the stream network and then routed farther downstream. The ISRRFL flag was added to the existing surface-water flags after the delivery-related flags and before the water-rights related flags. That is, it is inserted between flags IRDFL and IALLOT.

The use of the new Semi-Routed Runoff Returnflow flag, ISRRFL, is explained as follows (also, see tables 1 and 2):

ISRRFL = 0:   No locations along the stream network are specified for any farm, where semi-routed runoff returnflow is recharged into the stream network. Runoff is either automatically prorated over non-diversion-segment reaches located within a farm, or automatically discharged into one non-diversion-segment reach nearest to the lowest elevation of the farm.

ISRRFL = 1 or 2: For each farm, a location is specified anywhere along the stream network where semi-routed runoff returnflow is discharged anywhere in the active model domain. A farm-related list of row and column coordinates or segment and reach numbers for a point of runoff returnflow discharged will be read (only if "SFR" is specified in Name File).

ISRRFL = 1:   List of row and column coordinates or segment and reach numbers is read for the entire simulation.

ISRRFL = 2:   List of row and column coordinates or segment and reach numbers is read for each stress period.

If ISRRFL = 0, FMP2 attempts, by first priority, to automatically detect reaches of either non-diversion segments (if IRRFL = 1) or segments of any type (IRRFL = –1) located within a farm. For reaches of these segments that are located within a farm, the surface-water runoff from a farm is prorated over these reaches weighted by the stream lengths in each reach. We define this form of returnflow as "automatic fully-routed returnflow." If none of these segments are found within the farm's domain, then, by second priority, FMP2 automatically locates one remote reach of either a non-diversion segment (if IRRFL = 1) or of a segment of any type (IRRFL = –1) that is situated nearest to the lowest elevation of the farm. If such a remote reach is detected automatically, then the total runoff from a respective farm is discharged into this reach. This form of returnflow is called "automatic semi-routed returnflow." Reaches receiving the new "specified semi-routed returnflow" can be located on any type of segment. Notice that multiple farms may discharge into the same runoff returnflow reach as may be typical in many agricultural settings with elaborate drainage networks.

A list of row and column coordinates and segment and reach numbers of a returnflow reach has to be specified to receive semi-routed returnflow from a specific farm. This farm-specific list is read as a new Item 20b or 34b after previous Item 20a or 34a (list of coordinates for semi-routed deliveries) depending on whether these data are entered for the entire simulation or for each stress period. If ISRRFL ≥ 1 and a semi-routed runoff returnflow location is specified for a particular farm, then the automatic search for potentially existing reaches located within the farms that could receive automatic fully-routed runoff returnflow is disabled. For a farm where no runoff returnflow location is known, zero coordinates and zero segment and reach numbers have to be entered. If ISRRFL ≥ 1 and zero coordinates are specified for the returnflow location of a particular farm, then FMP2 applies either "automatic fully-routed runoff returnflow" or

"automatic semi-routed runoff returnflow" for that particular farm as described above for all farms if ISRRFL = 0. Setting ISRRFL ≥ 1 allows the user to be in control of manually specifying returnflow locations for farms where they are known, while FMP2 still automatically determines returnflow locations for other farms where they are unknown This ensures that simulated non-consumptive losses from a farm to surface-water runoff are accounted as returnflow from the respective farm to the stream network. This avoids mass loss to an open system and helps to account for 'all of the water everywhere and all of the time.' Simulating runoff outflows from farms without reallocating them back into the stream network would require the assumption that the runoff leaves the model domain. In FMP2, this assumption applies only if "SFR" is not specified in the MF2005 Name File.

The use of the new data list for Semi-Routed Runoff Returnflow locations as a new Item 20b or 34b is explained as follows (also, see tables 1 and 2):

Farm-ID: Farm identity to which the parameter below are attributed

Row: Row number of point of recharge of runoff returnflow

Column: Column number of point of recharge of runoff returnflow

Segment: Number of stream segment, in which the runoff returnflow reach is located. This number must be equal to the segment number of the identical stream reach as specified in column four of the data list in the SFR2 input file for the entire simulation.

Reach: Number of reach within a segment, into which the runoff returnflow is recharged. This number must be equal to the sequential number of a reach within a particular stream segment as specified in column five of the data list in the SFR2 input file for the entire simulation.

Four options of data input are available in order to uniquely identify the point of recharge of runoff returnflow within a cell:

Row	Column	Segment	Reach	Comments	
x	x	x	x	Full set of information is available	Maximum information
x	x	x	—	If more than one segment passes through the cell	Identification of location by coordinates
x	x	—	—	If just one segment passes through the cell	
0	0	x	x	If more than one segment passes through the cell	Identification of location by segment and reach numbers
[x = data input; 0 = input of zero; __ = no input or zero input]					

Farm Budget Output Options

A new output option called the Farm Budget required a new Farm Budget print flag, IFBPFL, specified in Item 2 of the FMP input data set. The Farm Budget is a budget of all physical flows into and out of a farm. It has to be viewed as a 'reference interface' on the ground surface rather than as a 'reference volume.' The present version of FMP2 does not include soil water storage or on-farm water storage in the Farm Budget.

The new mandatory Farm Budget print flag IFBPFL has to be entered in Item 2 after mandatory flag ISDPFL and before the optional flags {IOPFL}. The IFBPFL flag allows printing either a compact or detailed farm budget of flows into and out of each farm either to respective ASCII or binary files. "FB_COMPACT.OUT" or "FB_DETAILS.OUT" or to binary files. The compact budget summarizes various detailed individual components that are listed separately by the detailed budget. For instance, in the compact budget surface-water deliveries are listed cumulatively as Q-sw-in, and in the detailed budget they are listed as non-routed deliveries (Q-nrd-in), semi-routed deliveries (Q-srd-in), and routed deliveries (Q-rd-in). The budgets are printed for each farm and time step both as flow rates and as cumulative volumes. The use of the IFBPFL flag and budget components associated with either the compact or the detailed budget are listed in the "Data Input Instructions" section of this report. The budget components of the compact and detailed budgets (eq. 1 and 2) are defined in the "Output Data for Farm Process" section of this report.

Compact Farm Budget:

$$Q_p^{in} + Q_{sw}^{in} + Q_{gw}^{in} + Q_{ext}^{in} = Q_{et}^{out} + Q_{ineff}^{out} + Q_{sw}^{out} + Q_{gw}^{out} . \quad (1)$$

Detailed Farm Budget:

$$
\begin{aligned}
Q_p^{in} &+ Q_{nrd}^{in} + Q_{srd}^{in} + Q_{rd}^{in} \\
&+ Q_{wells}^{in} + Q_{e_{gw}}^{in} + Q_{t_{gw}}^{in} + Q_{ext}^{in} \\
= Q_{e_i}^{out} &+ Q_{e_p}^{out} + Q_{e_{gw}}^{out} + Q_{t_i}^{out} + Q_{t_p}^{out} + Q_{t_{gw}}^{out} + Q_{run}^{out} \\
&+ Q_{dp}^{out} + Q_{nrd}^{out} + Q_{srd}^{out} + Q_{rd}^{out} + Q_{wells}^{out} . \quad (2)
\end{aligned}
$$

Routing Information Output Options

A new optional Routing Information Print Flag, IRTPFL, is specified in Item 2 of the FMP input data set to provide, for each farm, information for both the routing of deliveries to the farm and for runoff returnflow from the farm. If IRTPFL is specified, this information will be written to a separate ASCII file, called ROUT.OUT. Otherwise, the information will be written to the listing file.

The output information regarding the architecture of the delivery system informs the user whether the farm can potentially receive either:

(a) fully-routed deliveries from the first, farthest upstream located reach of a sequence of automatically detected delivery-segment reaches within a farm, or whether

(b) semi-routed deliveries from specified stream reaches.

The output information regarding the architecture of the returnflow system informs the user whether potential runoff from the farm is returned as either:

(a) fully-routed to automatically detected returnflow-segment reaches within the farm, over which the runoff returnflow is prorated, weighted by the length of each reach, or

(b) semi-routed to specified stream reaches, or,

(c) in lieu of the first two options, semi-routed to automatically detected returnflow-segment reach nearest to the lowest elevation or a farm.

The format, in which the routing information is written, can be obtained from the chapter "Output Data for Farm Process."

Auxiliary Variable NOCIRNOQ

A new option that limits farm pumpage only to wells where an irrigation requirement exists required a new auxiliary flag AUX NOCIRNOQ specified in Item 2 of the FMP input data set. The specification of the optional flag AUX NOCIRNOQ for {flags for auxiliary variables} in Item 2 will prompt FMP2 to limit the distribution of farm pumpage to farm wells, whose row and column coincides with an uppermost-layer cell with a current irrigation requirement from active crops. NOCIRNOQ stands for "no crop irrigation requirement (CIR), no pumping (Q)." This feature is implemented by setting the maximum capacity of select farm wells to zero if, during a particular time step, no crop irrigation requirement occurs in a cell of the uppermost layer. At each new time step, the maximum capacity of each select well is reset to the default value. If some wells of a farm are deactivated, the remaining active wells will accordingly receive a higher demand to satisfy the pumping requirement.

The optional flag "AUX NOCIRNOQ" requires FMP2 to read an auxiliary variable [xyz] after the QMAXfact or QMAX variable of the farm wells list in Item 4 or 22, or after any other preceding auxiliary variable (for example, AUX QMAXRESET). The auxiliary variable for AUX NOCIRNOQ is defined to be a binary parameter that specifies which wells are selected for the NOCIRNOQ option. If a "1" is read, then the maximum capacity of the respective well is set to zero if, during a particular time step, no crop irrigation requirement of the top layer cell exists. At each new time step the maximum capacity of each select well will be reset to the default value.

Optional Flag WELLFIELD

A new option that allows a series of farms to receive their cumulative irrigation demand as simulated non-routed deliveries from well fields simulated as virtual farms required a new flag WELLFIELD specified in Item 2 of the FMP input data set for {flags for options}. This new option allows virtual farms with one or several wells (well fields) to receive a cumulative pumping requirement equal to the cumulative irrigation delivery requirement of other irrigated farms that are supplied by the well field. If the cumulative demand exceeds the cumulative maximum capacity of the well field, then other well fields can supply the residual demand. The cumulative pumpage of the well field that may be equal to or less than the desired demand will then be redistributed to the farms weighted by the total delivery requirement of the receiving farms (as $NRD_{i,j}$ to farm i receiving water from the first priority well field, eq. 3) or by the residual delivery requirement of the receiving farms (as $NRD_{i,j}$ to farm i receiving water from lower priority well fields, j, eq. 4). FMP2 then applies this redistributed rate as simulated non-routed deliveries to the respective farms.

$$NRD_{i,1} = \sum_{w=1}^{nw} Q_{w,1} \cdot \frac{TFDR_{i,1}}{\sum_{i=1}^{nf}\left(TFDR_{i,1}\right)}, \quad \text{for } j = 1. \qquad (3)$$

$$NRD_{i,j} = \sum_{w=1}^{nw} Q_{w,j} \cdot \frac{\left(TFDR_{i,j} - \sum_{j=1}^{j-1} NRD_{i,j}\right)}{\sum_{i=1}^{nf}\left(TFDR_{i,j} - \sum_{j=1}^{j-1} NRD_{i,j}\right)}, \text{for } j > 1, \qquad (4)$$

where

$i = 1, 2, ..., nf$ (number of farms receiving water from a well field),

$j = 1, 2,..., nm$ (number and priority of well fields, lower number = higher priority),

$w = 1, 2, ..., nw$ (number of wells associated with a well field),

$NRD_{i,j}$ is non-routed delivery to a particular farm i from a particular well field j,

$TFDR_{i,j}$ is total farm delivery requirement of a particular farm i receiving water from well field j,

$Q_{w,j}$ is maximum pumpage of well w associated with a particular well field j.

For farms that receive water from a particular well field, in Item 33, the NRDU flag has to be set to "minus the Farm ID of the virtual farm that contains the well field." For the virtual farm that contains the first (or second, or third, and so on) priority well field, the NRDU flag has to be equal to 1, listed as the third parameter of the set of three parameters for type 1 (or 2, or 3, and so on). Each well field represents a rank. Although, in FMP2, specified non-routed deliveries of a particular type (1, 2, 3, and so on) can be of any rank., the simulated "non-routed" delivery types that represent a well field have to be ordered in decreasing rank with increasing types.

In the example below, farms 1 through 4 receive water by first priority from the well field of virtual farm 5 and, if a residual delivery requirement remains, by second priority from virtual farm 6.

Farm ID	Unranked Non-Routed Delivery Type 1			Unranked Non-Routed Delivery Type 2			Unranked Non-Routed Delivery Type t_u
	NRDV Non-Routed Delivery [L³]	NRDR Priority Rank of NRD	NRDU Use-Flag	NRDV Non-Routed Delivery [L³]	NRDR Priority Rank of NRD	NRDU Use-Flag	...
1	0	1	−5	0	2	−6	
2	0	1	−5	0	2	−6	
3	0	1	−5	0	2	−6	
4	0	1	−5	0	2	−6	
5	0	1	1	0	2	0	
6	0	1	0	0	2	1	
[Notations of acronyms see chapter Data Input Instructions for FMP1 and New FMP2 Features under Non-Routed Surface Water Delivery — Farm-Related Data List (Item 33) and in the context of the wellfield option under "Options \ WELLFIELD"]							

A particular application can be the supply of water from an aquifer-storage-and-recovery system (ASR). The target percolation of the percolation pond of the ASR can be simulated as a "derived" irrigation demand of a virtual crop that is based on the known maximum infiltration rate of the ASR. The demand of the pond or, alternatively, injection wells can be supplied from whatever source the modeler intends to use (for example, semi-routed deliveries from the stream segment of a diversion or non-routed deliveries). The supply

from a recovery well or well field surrounding the ASR and from other second and third priority well fields to respective farms can be simulated by the WELLFIELD option. Hanson and others, 2008b described the use of the WELLFIELD option in the Pajaro Valley, California, to simulate the supply of the cumulative demand of farms connected to a coastal distribution system by a series of ranked well field (1st priority: recovery well of ASR, 2nd priority: blend wells, 3rd priority: municipal well field) (fig. 3).

FARM PROCESS SIMULATION OF PROJECT OPERATIONS

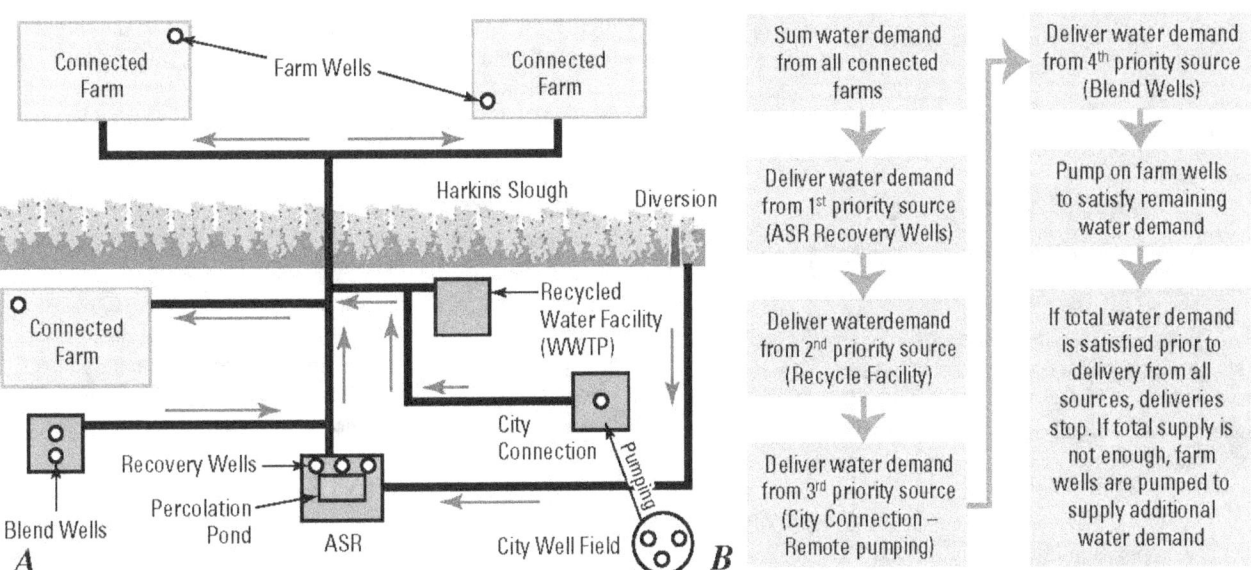

Figure 3. (A) Structure of local deliveries and flow chart showing (B) the order of operation of the simulation scheme for deliveries from an Aquifer-Storage-and-Recovery System (ASR) to regions serviced by the Coastal Distribution System, Pajaro Valley, California (Hanson and others, 2008b).

Optional Flag RECOMP_Q_BD

A new option is offered in FMP2 that allows the recomputation of flow terms at the end of each time step loop based on the simulated head of solution and dependent flows. This option required a new flag RECOMP_Q_BD specified in Item 2 of the FMP input data set for {flags for options}.

Head-dependent flow terms that depend on the head of the previous iteration before closure are generally recomputed in some packages (for example, SFR) for the head of the last iteration in the budget subroutines. This results in repetition of the solutions in the budget subroutines and may result in inconsistencies in the estimations of flows used by other features.

The calculation of a flow term that depends on the head of the previous iteration, $Q(h^{k-1})$, within the iteration loop can be repeated for simple calculations as a flow term that depends on the "head of solution," $Q(h^k)$, in the respective flow-budget (BD) routine within the time step loop (for example: stream leakage in the SFR Package). However, more complicated flow computations of MODFLOW packages or processes linked to other packages may not only depend on the head of solution but also on the flow term of the other linked package and may be better implemented as a separate routine outside the flow-budget (BD) routine (for example: net recharge or farm-well pumping in the Farm Process dependent on flow terms of the SFR or MNW Packages). A repetition of the call of the Farm Process FM-routine within the time step loop of

MF2005 was determined to improve computation speed and accuracy of the groundwater budget. However, the problem remains is that one of the two linked packages yielding a flow term will not be a function of the "flow of solution" of the other package unless the recomputation is repeated at the end of each time step loop. Here is an example for two linked packages (call to Package 1 from MF2005-main ahead of call to Package 2):

In the iteration loop of MF2005:

$$Q_1^k = Q_1(h^{k-1}, Q_2^{k-1})$$
$$Q_2^k = Q_2(h^{k-1}, Q_1^k)$$
with k = last iteration of solution in FM.

At the end of time step loop of MF2005:

$$Q_2^{\text{solution}} = Q_2\left(h^k, Q_1^k\right)$$
$$Q_1^{\text{solution}} = Q_1\left(h^k, Q_2^{\text{solution}}\right).$$

With respect to the sequence of placing the calls of the two packages, the recomputation at the end of the time step routine has to be reversed. Q_2 was called last during the last iteration (of solution). It therefore depends already on the Q_1 of solution. Hence, at the end of the time step loop, Q_2 should probably be computed first, and Q_1 second.

In FMP2, when specifying the option RECOMP_Q_BD, the re-computation of the Farm Process FM-routine is invoked at the end each time step loop.

Optional Flag MNWCLOSE

If FMP farm wells are linked to the MNW Package (indicated by negative farm-well ID), then farm-well pumping requirements determined by FMP2 are passed on to the MNW Package as desired pumping rates. In the next iteration, the MNW Package passes the actual pumping rates back to FMP2. During this iteration, FMP2 already calculates a new, slightly modified, pumping requirement on the basis of a slightly changed demand that depends on slightly changed aquifer heads. If the actual possible pumping rate in the MNW Package is bounded by a hydraulically found maximum possible rate (for example, by reaching a head or drawdown constraint), then such an actual pumping rate is passed back to FMP2 as a new, simulated, maximum capacity that replaces

the maximum capacity initially specified in the FMP input data set. In that case, the pumping requirement in FMP2 cannot exceed the hydraulically found maximum capacity in MNW. That is, as long as the pumping requirement is not bounded by a simulated maximum possible pumping rate in MNW, the pumping requirement calculated in FMP2 and the actual pumping rate determined in MNW have to converge. As both rates depend on the aquifer head, the convergence of the FMP pumping requirement and the MNW actual pumping rate beyond a certain flow closure criterion may require more iterations by lowering head or residual closure criteria as specified in MF2005 solver packages.

A new optional flag MNWCLOSE is specified in Item 2 of the FMP input data set for {options} to indicate if the head- and residual-closure criteria of the MODFLOW solver package should be adjusted to allow convergence of the FMP pumping requirement to pumping simulated by the linked MNW Package. If MNWCLOSE is specified, then new parameters QCLOSE, HPCT, and RPCT are specified in Item 2 after MNWCLOSE.

When specifying the optional flag MNWCLOSE in Item 2, the MF2005-FMP2 will adjust the preset closure criteria of the solver data input file (for example, HCLOSE and RCLOSE in PCG7) to allow convergence of the desired pumpage (FMP pumping requirement) to the actual pumpage simulated by the MNW Package linked to FMP2. Wells simulated only by MNW (not linked to FMP2) are exempt from this check. As long as a so-called QCLOSE closure criterion set after flag MNWCLOSE is not met, the iteration loop will be prevented from solving, even if HCLOSE and RCLOSE were met during the current iteration. Aside from QCLOSE, other input requirements are fractions HPCT and RPCT, by which either the Head Change Closure Criterion or the Residual Change Closure Criterion has to be reduced.

Matrix of On-Farm Efficiencies

In addition to specifying one value of on-farm efficiency per farm, now a matrix of efficiencies for any farm and any crop can be specified (rows: farm ID; columns: crop type ID), which required modifications to parameter OFE specified in Items 7 or 24 of the FMP input data set. The user may also specify efficiencies varying from crop to crop for some farms while still specifying only one efficiency value (that is, not varying by crop) for some other farms. In the latter case, the efficiency entered in column 2 of Item 7 or Item 24 (column 1 = farm ID) is assumed to be valid for all other crops and the matrix fields do not have to be filled for other crops.

Input Requirements

For a matrix of on-farm efficiencies, the required data input in Item 7 or 24 is as follows (see tables 1 and 2):

[FID OFE(FID,CID=1), OFE(FID,CID=2), ... , OFE(FID,CID=NCROPS)] read NFARMS times for each simulation if IEFFL = 1 (Item 7) or for each stress period if IEFFL = 2 (Item 24). (FID= Farm ID; CID = Crop-ID)

In matrix form (NFARMS = number of farms; NCROPS = number of crops):

	Column 1	Column 2	Column 3	. . .	Column NCROPS+1
Row 1	1	$OFE_{1,1}$	$OFE_{1,2}$	...	$OFE_{1,NCROPS}$
Row 2	2	$OFE_{2,1}$	$OFE_{2,2}$	...	$OFE_{2,NCROPS}$
. . .	...	...	...	...	...
Row NFARMS	NFARMS	$OFE_{NFARMS,1}$	$OFE_{NFARMS,2}$		$OFE_{NFARMS,NCROPS}$

CAUTION: Comments for each farm can be entered neither to the right of efficiencies specified by crop nor to the right of just one value per farm in column 2. Should the user enter a comment, the input error will be printed to the list file: ERROR CONVERTING "..." TO A DP-REAL NUMBER IN LINE ... .

Data Output

For each farm, an output "composite efficiency" will be printed together with the farm demand and supply budget for each iteration, each time step, or selected time steps either to the list file, to an ASCII file called FDS.OUT, or to a binary file as specified by the Farm Supply and Demand Print Flag, ISDPFL. This "composite efficiency" is an area-weighted average of either specified efficiency values (IEBFL = 1) or simulated head-dependent efficiencies (IEBFL = 2, 3) of all model cells in a farm, weighted by the area of each cell.

Non-Irrigated Crops

Non-irrigated crops are designated by a new parameter NONIRR set equal to one in either column 12 of the data list in Item 15 of the FMP input data set if ICUFL = 3 (table 1) or column 3 of the data list in Item 27a of the FMP input data set if ICUFL ≤ 2 (table 2). This depends on whether the crops' consumptive use is derived from data specified either for the entire simulation or for each stress period.

If ICUFL = 3, then a list of crop specific parameters (Item 15) and a time-series of climate data (Item 16) are specified for the entire simulation (Schmid and others 2006, pages 47 and 48), which allows the FMP-internal derivation of a potential crop evapotranspiration flux value for each crop

and time step. In this case, non-irrigated crops are designated for the entire simulation by NONIRR set equal to one in column 12 of the list of crop specific parameters in Item 15.

If ICUFL = 1 or 2, then a crop-specific list of potential crop evapotranspiration flux values is specified for each stress period in column 2 of the data list in Item 27a. If ICUFL = −1, then a crop specific list of crop coefficients is specified for each stress period in column 2 of the data list in Item 27a, followed by a constant or array of reference evapotranspiration in Item 27b for each stress period, which allows the derivation of a potential crop evapotranspiration flux value as the product of crop coefficients and reference evapotranspiration. For both cases (ICUFL = 1,2 or ICUFL = −1), non-irrigated crops are designated for each stress period by NONIRR set equal to one in column 3 of the data list in Item 27a.

For irrigated crops, the new parameter NONIRR is set to zero or no data entry is required in column 3 of Item 27a (ICUFL ≤ 2), or in column 12 of Item 15 (ICUFL = 3), respectively.

FMP2 does not calculate an irrigation requirement or excess irrigation returnflows for non-irrigated crops. However, it does account for transpiration and evaporation portions supplied by precipitation and groundwater uptake, as well as for excess precipitation runoff returnflows and deep percolation.

For Non-irrigated crops, the required data input for the new flag NONIRR in Item 27a or in Item 15 is as follows (see tables 1 and 2):

[Crop-ID CU NONIRR] read NCROPS times if ICUFL = –1, 1, or 2 (Item 27a)

	Column 1	Column 2	Column 3
Row 1	1	CU_1	$NONIRR_1$
Row 2	2	CU_2	$NONIRR_2$
. . .	...	...	...
Row NCROPS	NCROPS	CU_{NCROPS}	$NONIRR_{NCROPS}$

Or

[Crop-ID BaseT MinCutT MaxCutT C0 C1 C2 C3 BegRootD MaxRootD RootGC NONIRR] read NCROPS times if IRTFL = 3, or ICUFL = 3, or IPFL = 3 (Item 15)

	Column 1	Column 2	Columns 3 to 11	Column 12
Row 1	1	$BaseT_1$	...	$NONIRR_1$
Row 2	2	$BaseT_2$	...	$NONIRR_2$
. . .	...	...	...	
Row NCROPS	NCROPS	$BaseT_{NCROPS}$	...	$NONIRR_{NCROPS}$

Root Uptake under Variably Saturated Conditions

In FMP1, the simulation of the transpiration of natural vegetation or crops was limited to uptake from root zones under unsaturated conditions. A water level rising into the root zone causes saturated conditions, under which the stress response of natural vegetation or crops becomes zero. A new concept was developed for FMP2 that allows the simulation of natural vegetation or crops (for example, rice and willow trees) that do not reduce their uptake as a result of anoxic conditions in the unsaturated zone and maintain maximum uptake even under saturated conditions until, eventually, they reduce their uptake as positive pressure heads increase. The first section briefly reviews the previous concept in FMP1 of root uptake under unsaturated conditions. The review of the previous concept refers to text and graphs in the FMP1 user guide (Schmid and others, 2006). The second section elaborates on the new concept in FMP2 of root uptake under variably saturated conditions. The third section discusses changes to the input requirements for pressure heads that define crop-specific stress response functions.

Previous Concept of Root Uptake under Unsaturated Conditions in FMP1

FMP assumes that the actual transpiration of crops is reduced proportionally to the reduction of the total root zone (TRZ) to an active root zone by wilting and anoxia. Prior to the current FMP2 modifications, this concept extended only to crops, whose stress response to water does not allow any uptake under saturated conditions.

The conceptualization of root uptake from an unsaturated root zone in FMP is described in detail in Schmid and others (2006, pages 11, 15, 46, and 76) and in Schmid (2004, pages 80, 83, 84, and 86). Varying hydraulic pressure heads within a root zone impose different levels of stress on a crop resulting in water uptake ranging between a maximum and zero. The functionality between dimensionless water uptake, α, $(0 \leq \alpha \leq 1)$ and pressure head, ψ, is called a 'water stress response function.' Such a crop-specific water stress response function can be defined by four negative critical pressure heads at which water uptake ceases as a result of either anoxia or wilting (ψ_1, ψ_2) or at which water uptake is at its maximum (ψ_1, ψ_2).

Among several sources in the literature, Taylor and Ashcroft (1972) and Wessling (1991) provide the most detailed databases for stress response pressure heads for numerous crops. If data are lacking for aerobic crops, ψ_1 may be approximated as the air entry pressure head ψ_a, in the water retention function, where the water content θ approaches the porosity n. Common bounds for the field capacity, if known, may provide an approximation of the value for ψ_2 (normally between –0.06 and –0.3 bar or –60 and –300 cm pressure head). A maximum allowable depletion (in percent) describes the reduction of the water content at field capacity. The according minimum allowable water content, below which transpiration is reduced, can be related back to a pressure head to approximate the value of ψ_3. The permanent wilting point for most crops, ψ_w, is at about –15 bar or –15,000 cm pressure head and can be used as an indication of the ψ_4 value. However, the approximation of ψ_1 though ψ_4 from the air entry pressure head, a range of field capacity pressure heads, the maximum allowable depletion, and the permanent wilting point is problematic. While the air entry pressure head and the field capacity vary with soil type, the maximum allowable depletion and the permanent wilting point vary with crop type. Because FMP requires stress response pressure heads to be crop-type specific attributes, the user is encouraged to search the literature for databases of strictly crop-type related stress response pressure heads.

FMP simplifies the stress response function to a step function, where water uptake is considered at maximum between the averages of ψ_1 and ψ_2, and of ψ_3 and ψ_4 (Schmid and others, 2006, fig. 8A, p.15). These averages are then compared with pressure heads found by an analytical solution of the vertical pressure head configuration across the root zone (Schmid and others, 2006, fig. 8B, p.15). In FMP, regions of the root zone with negative pressure heads smaller than the average of ψ_4 and ψ_3 or greater than the average of ψ_2 and ψ_1 are considered inactive wilting and anoxia zones, respectively (WZ, AZ) (Schmid and others, 2006, fig. 8B, p.15). For a water level at the bottom of the root zone (h_b), the residual active unsaturated root zone (AURZ) is equal to the TRZ minus WZ and AZ. As the groundwater level rises, the vertical pressure head distribution is shifted upward. As water levels rise, the WZ at the top end of the pressure head distribution gradually becomes eliminated and the active root zone remains constant until the water level reaches a point, where the depth of the WZ is zero (water level at that point = h_{w0}). For water levels rising beyond this point, the AURZ is reduced linearly until the top of the anoxia fringe above the water level reaches the ground-surface elevation (GSE). At this position of the water level, transpiration reaches extinction (water level at that point = h_{ux}).

$$AURZ(h) = \begin{array}{ll} TRZ - AZ - WZ, & h_b \geq h > h_{w0} \\ GSE - AZ - h, & h_{w0} \geq h > h_{ux} \\ 0 & h > h_w . \end{array} \qquad (5)$$

For water levels at or above the bottom of the root zone, the root uptake from groundwater under unsaturated conditions can be formulated as:

$$T_{gw-act-unsat}(h) = T_{c-pot} \cdot AURZ(h) / TRZ, \qquad (6)$$

where

T_{c-pot} is potential transpiration;

$T_{gw-act-unsat}$ is actual transpiration (root uptake) from unsaturated root zone supplied by capillary rise from groundwater;

$AURZ$ is active unsaturated root zone;

TRZ is total root zone; and

(h) is head dependent (function of groundwater level).

Expanded Concept of Root Uptake under Variably Saturated Conditions

An expansion of the previous concept of root uptake under variably saturated conditions was needed, because certain crops and riparian vegetation (for example, rice and willow trees) do not reduce their uptake because of anoxic conditions in the unsaturated zone. However, they do eventually reduce their uptake as positive pressure heads increase in the saturated root zone or, for ponding conditions, up to a user-specified limit of water level above the ground-surface elevation.

For deep root zones and for groundwater levels ranging within the root zone, particular cases are possible, where uptake under both unsaturated and saturated conditions occurs. Above the groundwater level, zero or full uptake may occur under unsaturated conditions within the WZ and the AURZ, respectively. For crops characterized by positive critical pressure heads ψ_1 and ψ_2, the AURZ is not restricted by anoxia (AZ = 0) (fig. 4, above groundwater level). Below the groundwater level, full or reduced uptake may occur under saturated conditions within the active saturated root zone (fig. 4, below groundwater level):

- Full uptake under saturated conditions occurs for a region of positive pressure heads within the root zone ranging between zero at the groundwater level and the user specified pressure head ψ_2. This region of the root zone is defined as Active Saturated Root Zone 1, or ASRZ1. Within this zone, the stress response to water uptake, α, is equal to 1, indicating that full uptake is possible. For water levels rising above the GSE (not displayed in fig. 4), the ASRZ1 extends from the GSE to where the critical pressure head ψ_2 is found.

- Reduced uptake under saturated conditions occurs for a region of positive pressure heads ranging between ψ_2 (full uptake) and the lesser of ψ_1 (zero uptake) and the pressure head at the bottom of the root zone. We define this region of the root zone as Active Saturated Root Zone 2, or ASRZ2. Within this zone, the stress response to water uptake, α, is taken to be equal to the average of stress responses, $\overline{\alpha}$, owing to pressure heads that are found within the root zone between ψ_2 and the lesser of ψ_1 and the pressure head at the bottom of the root zone. Where ψ_1 is found below the bottom of the root zone (fig. 4), ASRZ2 is not bound by ψ_1 but by a nonzero pressure head at the bottom of the root zone.

For water levels at and above the bottom of the root zone, the uptake under saturated conditions is formulated as:

$$T_{gw-act-sat}(h) = T_{c-pot} \cdot (ASRZ1(h) + ASRZ2(h) \cdot \overline{\alpha}(h))/TRZ. \quad (7)$$

For water levels at and above the bottom of the root zone, the total uptake is formulated as:

$$T_{gw-act}(h) = T_{c-pot} \cdot (AURZ(h) + ASRZ1(h) + ASRZ2(h) \cdot \overline{\alpha}(h))/TRZ, \quad (8)$$

where

T_{c-pot} is potential transpiration;

$T_{gw-act-sat}$ is portion of actual transpiration (root uptake) from active saturated root zone below the groundwater level;

T_{gw-act} is total actual transpiration (root uptake) from active saturated root zone below the groundwater level and from active unsaturated root zone above the groundwater level;

$AURZ$ is active unsaturated root zone;

$ASRZ1$ is active saturated root zone with maximum uptake;

$ASRZ2$ is active saturated root zone with reduced update;

TRZ is total root zone;

$\overline{\alpha}$ is average of stress responses found in $ASRZ2$; and

(h) is head dependent (function of groundwater level).

Noticeably, ASRZ1, ASRZ2, and $\overline{\alpha}$ depend on the vertical location of the hydrostatic pressure heads ψ_1 and ψ_2. Because ψ_1 and ψ_2 move vertically up or down as the water level rises or falls, the terms ASRZ1, ASRZ2, and $\overline{\alpha}$ depend on the simulated groundwater level, and therefore, are head-dependent terms. To avoid the term $ASRZ2(h) \cdot \overline{\alpha}(h)$ becoming nonlinear in head, we evaluate on the basis of the head of the previous iteration (k–1), whereas ASRZ2 is related to the head of the current iteration (k) (notations see equations 7 and 8):

$$T_{gw-act}(h^k) = T_{c-pot} \cdot (AURZ(h^k) + ASRZ1(h^k) + ASRZ2(h^k) \cdot \overline{\alpha}(h^{k-1}))/TRZ. \quad (9)$$

Schmid and others (2006) explained how $T_{gw-act-unsat}$ can be split into non-head-dependent and head-dependent terms. Similarly, $T_{gw-act-unsat}$, is separated into terms either dependent or not dependent on the head of the current iteration. While figure 4 demonstrates a situation for a particular water-level elevation, figure 5 illustrates the conceptual approximation to the change of all transpiration and evaporation components with varying groundwater level (fig. 5, example 1). Noticeably, this report only discusses the transpiration and not the evaporation components, as the latter ones are explained already in the FMP1 user guide (Schmid and others, 2006) and were not changed in FMP2.

This concept allows the simulation of water uptake and irrigation requirements of natural vegetation or crops (for example, rice and willows) rooting in soils that are fully or partially saturated by the groundwater level rising into the root zone or even above ground surface (for example, in alluvial valleys). Under such conditions, irrigation is required only for vegetation specified as irrigated crops for special cases where uptake from groundwater does not fully satisfy the potentially possible transpiration.

Depending on where the water level is positioned (above, within, or below the root zone), this new FMP2 concept considers five different cases of combinations of up to four transpiration components. These components are fed either by capillary rise from groundwater (unsaturated root zone), by direct uptake from groundwater (saturated root zone), by irrigation, or by precipitation. For instance, for Case 3 (fig. 5, example 1, left margin), the water level rises only slightly above the bottom of the root zone and wilting still might occur in the drying top soil. Transpiration is fed by groundwater uptake from the unsaturated and saturated part of the root zone. The deficit between the transpiration from groundwater and the maximum possible transpiration may be supplemented by precipitation or irrigation.

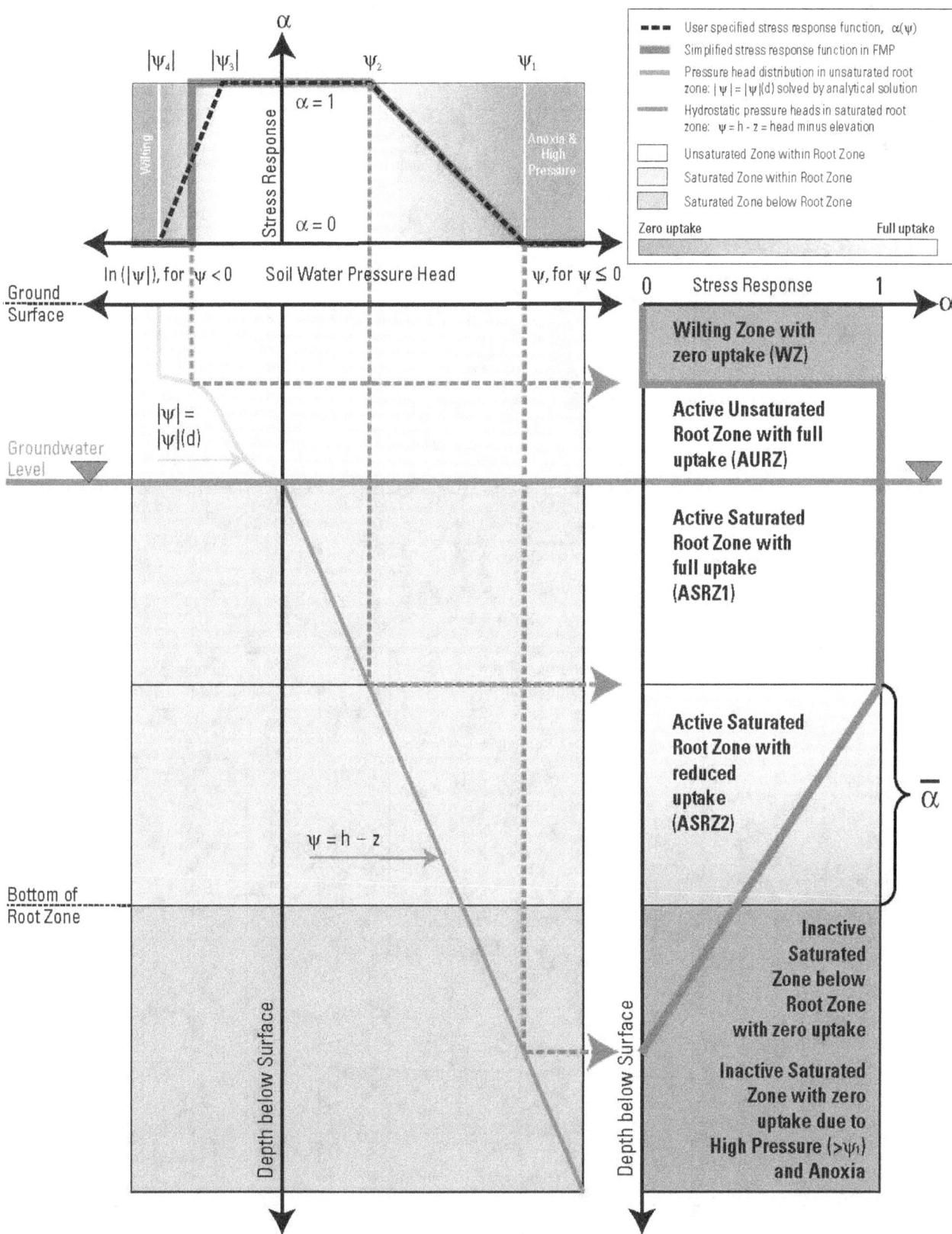

Figure 4. Evaluation of active and inactive portions of a variably saturated root zone in FMP2. (Concept of stress response, α as a function of pressure heads, ψ, varying over depth, d). (definition of WZ, AURZ, ASRZ1, ASRZ2: see equation 8; h = groundwater head, z = elevation.)

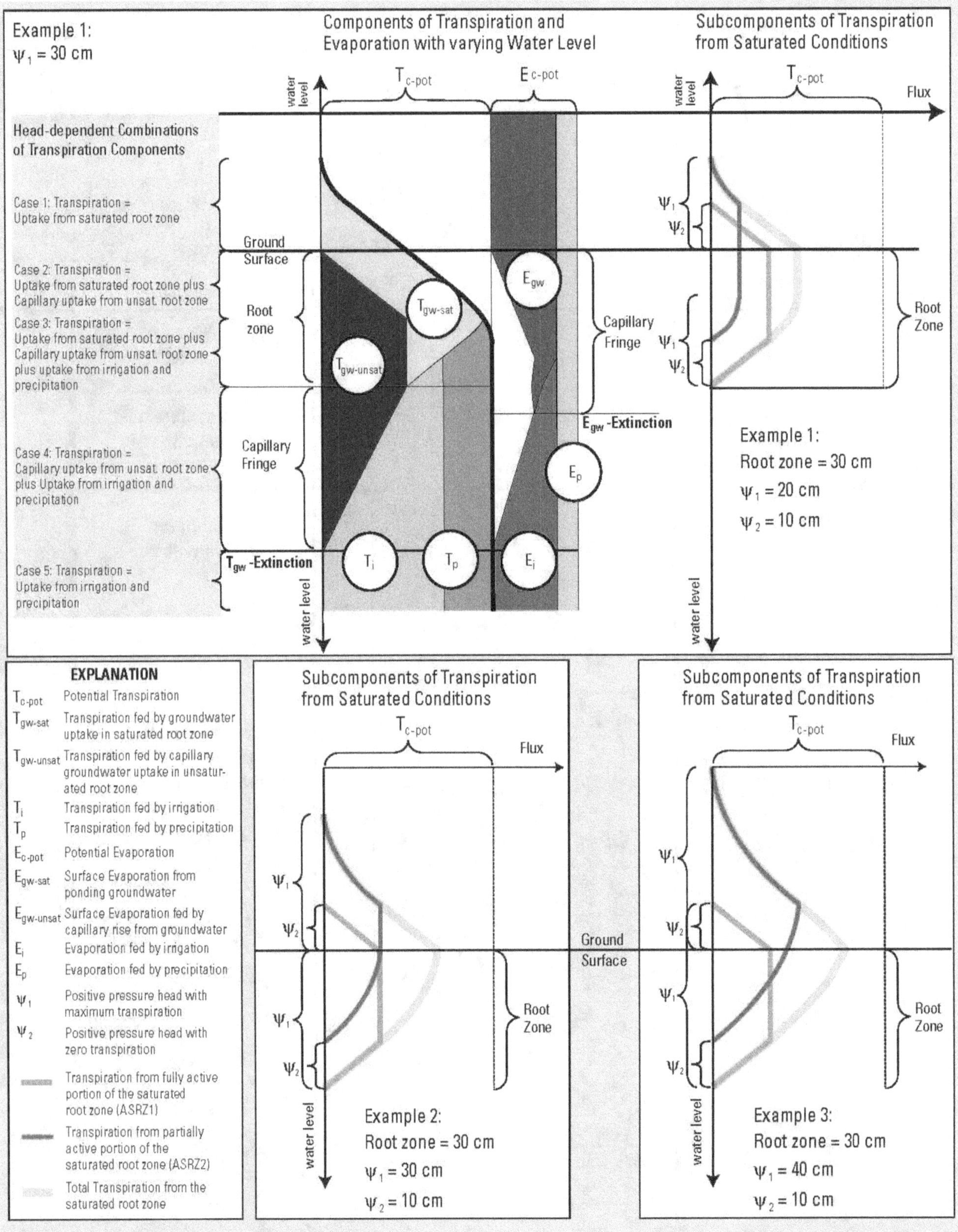

Figure 5.   Conceptualization to the change of transpiration uptake from a saturated root zone with varying water levels (three examples with different ψ_i values; example 1 at top includes conceptualization of all transpiration and evaporation components with varying water levels).

However, if the water level continues to rise (Case 2, fig. 5, example 1, left margin), all possible transpiration will occur by groundwater uptake from the unsaturated and saturated root zone. Finally, when the water level rises above the ground surface and ponding occurs, then uptake will only take place from the saturated, inundated root zone (case 1, fig. 5, example 1, left margin).

Examples 1, 2, and 3 (fig. 5) show how the total transpiration uptake from the saturated root zone (light green curve) is composed of the uptake from the fully active and partially active portion of the saturated root zone. The uptake from the fully active root zone (light blue curve) is a piecewise linear approximation. The uptake from the partially active root zone (purple curve) depends on the product of two head-dependent terms: the depth of this zone and the average stress response, $\bar{\alpha}$ within this zone. Therefore, as shown in figure 5, this portion of the uptake is nonlinear with changing head (eq. 8). For select positive ψ_1 and ψ_2 values, the range of positive pressure heads with reduced uptake ($\psi_1 - \psi_2$) may be less than the thickness of the total root zone. In this case, the "partial uptake zone," ASRZ2, and the average stress response, $\bar{\alpha}$, within that zone may remain constant with a moving water level, as long as the elevation where ψ_2 is found (head $-\psi_2$) is less than the ground-surface elevation, and as long as the elevation where ψ_1 is found (head $-\psi_2$) is greater than the bottom of the root zone.

The simulation of root uptake under variably saturated conditions requires modifications to the parameter ICUFL specified in Item 2 of the FMP input data set and parameters PSI(1), PSI(2), PSI(3), and PSI(4) in Item 14 of the FMP input data set. The use of FMP1 was limited to natural vegetation and crops that consume water only under unsaturated conditions. The original input requirements for crop-specific stress response functions required absolute values of negative pressure heads in the unsaturated root zone, between which root uptake is at maximum and at which uptake becomes zero due to wilting and anoxia.

Input Requirements

The new, expanded concept of consumptive use in FMP2 allows the simulation of natural vegetation or crops that take up water from portions of the variably saturated root zone with both negative and positive pressure heads. Therefore, the user must be able to specify any negative or positive pressure head at which root uptake is at maximum or zero. For saturation-intolerant natural vegetation or crops the user is required to specify negative pressure head values for the stress response function for unsaturated conditions. For natural vegetation or crops that tolerate saturated conditions, ψ_1 and ψ_2 have to be positive pressure heads. These values describe a linear decrease of uptake from optimal (at ψ_2) to zero (at ψ_1) due to anoxia and increasing pressure. The critical pressure heads ψ_3 and ψ_4 remain negative, since they represent the diminishing of uptake from maximum (at ψ_3) to zero (at ψ_4) due to wilting in drying, unsaturated conditions.

The input instructions for Item 14 (Schmid and others, 2006, p. 76) have changed to (see tables 1 and 2):

PSI(1): Negative or positive pressure head, at which root uptake becomes zero due to anoxia or high pressure [L]

PSI(2): Negative or positive pressure head, at which root uptake is at maximum and from which uptake decreases with rising pressure head due to anoxia [L]

PSI(3): Negative pressure head, at which root uptake is at maximum and from which uptake decreases with falling pressure head due to wilting [L]

PSI(4): Negative pressure head, at which root uptake becomes zero due to wilting [L]

Data Input Instructions for FMP1 and New FMP2 Features

The summary of data input parameters (tables 1 and 2) includes the previous unchanged FMP1 and changed or new FMP2 input items. The position of changed or new items within the previous FMP1-numbering scheme of data input items is highlighted in yellow. Flags or parameters highlighted in blue text represent modified features from FMP1 and flags or parameters highlighted in red text represent new features of FMP2 not previously available in FMP1.

Data input for FMP2 is read from the file designated as type "FMP" in the name file. This chapter contains two sections. The first section describes the data input requirements for each input item. An input item can consist of a comment, of flags, or of scalar-, list- or array-variables. Optional variables and optional flags are shown in brackets, [ ], and curly braces, { }, respectively. Two-dimensional arrays are listed together with their array dimensions (NCOL, NROW). Data lists or arrays, which are read by MF2005 or FMP2 utility modules, are indicated by "read*" and by a footnote that explains, which utility array readers are used, respectively. The reader is referred to the FMP1 user's guide (Schmid and others, 2006, p. 64 to 69) for a reference of each footnote to a respective utility read module and for further explanations of how information is read by the utility read modules, the input structure of the array and list reading utility modules, and how to apply scale factor multipliers to the variables. The second section provides an explanation of the fields itemized in the input instructions in the first section.

Input Data for FMP2

Data for each Simulation

Table 1 summarizes the data input for parameters the user may need to specify for the entire period of the simulation.

Data for each Stress Period

Table 2 summarizes the data input for parameters the user may need to specify for each stress period over the entire period of the simulation.

Explanation of Fields Used in the Input Instructions

Dimensions and Flags (Item 2)

Parameter Dimensions (Item 2)

NPFWL	Number of farm well parameters (changeable parameter is a multiplier of the maximum capacity).
MXL	Maximum number of parameter farm wells.
MXACTFW	Maximum number of active farm wells including parameter and nonparameter farm wells. Nonparameter farm wells are wells, whose maximum capacity is different for each stress period. In this case, each well-list (layer, location, farm-well farm ID, maximum capacity) would have to be read for each stress period. However, since the maximum capacity in most cases is thought to be constant for the entire simulation, usually the maximum number of nonparameter farm wells will be zero, that is, MXL = MXACTFW.
NFARMS	Number of farms.
NCROPS	Number of crop types.
NSOILS	Number of soil types.

Table 1.   Summary of FMP2 input for data required for the entire period of simulation.

[Yellow: Position of changed or new items within FMP1-numbering scheme. *Blue* text: Flags or parameters representing modified features from FMP1. Red text: Flags or parameters representing new features of FMP2]

Item No.	Input instruction for each item	
0	[#Text] read if '#' is specified (can be repeated multiple times)	
1	[*PARAMETER* NPFWL MXL] read if word 'PARAMETER' is specified	
2a	[*FLAG_BLOCKS*] specify word 'FLAG_BLOCKS' only if flags are to be specified by blocks	
2b	read flags from a single line if word 'FLAG_BLOCKS' is not specified in Item 2a: MXACTW NFARMS NCROPS NSOILS IRTFL *ICUFL* IPFL IFTEFL IIESWFL IEFFL IEBFL IROTFL IDEFFL {IBEN} {ICOST} *ICCFL* INRDFL {MXNRDT} *ISRDFL IRDFL* ISRRFL IRRFL IALLOT {PCLOSE} IFWLCB IFNRCB ISDPFL IFBPFL {IRTPFL} {IOPFL} {IPAPFL} {*Flags for Auxiliary Variables*} {*Flags for Options*} {QCLOSE HPCT RPCT}	
2c	read flags by blocks if word "FLAG_BLOCKS" is specified in Item 2a: MXACTW NFARMS NCROPS NSOILS IRTFL *ICUFL* IPFL IFTEFL IIESWFL IEFFL IEBFL IROTFL IDEFFL {IBEN} {ICOST} ICCFL INRDFL {MXNRDT} *ISRDFL IRDFL* ISRRFL IRRFL IALLOT {PCLOSE} IFWLCB IFNRCB ISDPFL IFBPFL {IRTPFL} {IOPFL} {IPAPFL} *Flags for Auxiliary Variables* *Flags for Options* {QCLOSE HPCT RPCT}	Dimensions When-to-read Flags Water Policy Flags Consumptive Use Concept Flag Surface-Water Flags Print Flags or Print Units
3	[PARNAM PARTYP PARVAL NLST] [*INSTANCES* NUMINST]	Repeat Items 3 combined with the indicataed repetitions of Item 4 NPFWL times if NPFWL > 1.
4a 4b	[INSTNAM] [Layer Row Column Farm-Well-ID Farm-ID QMAXfact] [*xyz*] After each Item 3 for which the keyboard "INSTANCES" **is not** entered, read Item 4b and not Item 4a. After each Item 3 for which the keyword "INSTANCES" **is** entered, read Item 4a and Item 4b for each instance. NLST repetitions of Item 4b are required; they are read by subroutine [3]. (SAFC of the utility subroutine [3] applies to QMAXfact). The NLST repetitions of Item 4b follow each repetition of Item 4a when PARNAM is time varying.	Items 3 and 4 are not read if NPFWL is 0. If PARNAM is to be a time-varying parameter, the keyword "INSTANCES" and a value for NUMINST must be entered.
5	GSURF(NCOL,NROW) read* with [2]	
6	FID(NCOL,NROW) read* with [1]	
7	[Farm-ID OFE] or [Farm-ID OFE(FID,CID), OFE(FID,CID), ... , OFE(FID,CID$_{NCROPS}$)] read* NFARMS times with [5] if IEFFL = 1	
8	SID(NCOL,NROW) read* with [1]	
9	Soil-ID CapFringe [A-Coeff B-Coeff C-Coeff D-Coeff E-Coeff], or Soil-ID CapFringe [Soil-Type] (parameters in brackets only if ICCFL = 1) read* NSOILS times with [6]	
10	[CID(NCOL,NROW)] read* with [1] if IROTFL $\geq$ 0	
11	[Crop-ID ROOT] read* NCROPS times with [4] if IRTFL = 1	
12	[Crop-ID FTR FEP FEI] read* NCROPS times with [5] if IFTEFL = 1	
13	[Crop-ID FIESWP FIESWI] read* NCROPS times with [5] if IIESWFL = 1	
14	[Crop-ID *PSI(1) PSI(2) PSI(3) PSI(4)*] read* NCROPS times with [5] if ICCFL = 1	
15	[Crop-ID BaseT MinCutT MaxCutT C_0 C_1 C_2 C_3 BegRootD MaxRootD RootGC {NONIRR}] read* NCROPS times with [5] if IRTFL = 3, or ICUFL = 3, or IPFL = 3	
16	[TimeSeriesStep MaxT MinT Precip ETref] read* LENSIM times with [5] if IRTFL = 3, or ICUFL = 3, or IPFL = 3 (LENSIM = length of simulation expressed as total number of time-series steps; length of time-series step defined by ITMUNI in the Discretization File)	
17	[Crop-ID IFALLOW] read* NCROPS times with [7] if IDEFFL = -2	
18	[Crop-ID WPF-Slope WPF-Int Crop-Price] read* NCROPS times with [5] if IDEFFL > 0 and if IBEN = 1	
19	[Farm-ID GWcost1 GWcost2 GWcost3 GWcost4 SWcost1 SWcost2 SWcost3 SWcost4] read* NFARMS times with [5] if IDEFFL > 0 and ICOST = 1	
20a	[Farm-ID Row Column Segment *Reach*] read* NFARMS times with [7] if ISRDFL = 1	
20b	[Farm-ID Row Column Segment Reach] read* NFARMS times with [7] if ISRRFL = 1	

Table 2.   Summary of FMP2 input for data required for each stress period over the entire period of simulation.

[Yellow: Position of changed or new items within FMP1-numbering scheme. *Blue* text: Flags or parameters representing modified features from FMP1. Red text: Flags or parameters representing new features of FMP2]

Item No.	Input instruction for each item
21	ITMP NP read
22	[Layer Row Column Farm-Well-ID Farm-ID QMAX] [*xyz*] read* ITMP times with [3] if ITMP > 0
23	[Pname] [Iname]] read NP times if NP > 0. Iem 23 is not read if NP is 0. Iname is read if Pname is a time-varying parameter.
24	[Farm-ID OFE] or [Farm-ID OFE(FID,CID), OFE(FID,CID), ... , OFE(FID,CID$_{NCROPS}$)] read* NFARMS times with [5] if IEFFL = 2
25	[CID(NCOL,NROW)] read* with [1] if IROTFL = -1
26	[Crop-ID ROOT] read* NCROPS times with [4] if IRTFL = 2
27a	[Crop-ID *CU* {NONIRR}] read* NCROPS times with [4] if ICUFL = 2
27b	ETR(NCOL,NROW) read with [2] if ICUFL = 1 or -1
28	[Crop-ID FTR FEP FEI] read* NCROPS times with [5] if IFTEFL = 2
29	[Crop-ID FIESWP FIESWI] read* NCROPS times with [5] if IIESWFL = 2
30	[PFLX(NROW,NCOL)] read* with [2] if IPFL = 2
31	[Crop-ID WPF-Slope WPF-Int Crop-Price] read* NCROPS times with [5] if IDEFFL > 0 and if IBEN = 2.
32	[Farm-ID GWcost1 GWcost2 GWcost3 GWcost4 SWcost1 SWcost2 SWcost3 SWcost4] read* NFARMS times with [5] if IDEFFL > 0 and ICOST = 2.
33	[Farm-ID (NRDV NRDR NRDU)$_1$, (NRDV NRDR NRDU)$_2$, ... , (NRDV NRDR NRDU)$_{MXNRDT}$] read* NFARMS times with [5] if INRDFL = 1. A maximum number of MXNRDT types of nonrouted deliveries is read for each farm. One set of variables NRDV, NRDR, and NRDU is read for a certain unranked type *t* of a nonrouted delivery by (NRDV NRDR NRDU)$_t$.
34a	[Farm-ID Row Column Segment *Reach*] read* NFARMS times with [7] if ISRDFL = 2
34b	[Farm-ID Row Column Segment Reach] read* NFARMS times with [7] if ISRRFL = 2
35	[ALLOT] read if IALLOT = 1
36	[Farm-ID CALL] read* NFARMS times with [5] if IALLOT = 2

'When-to-Read-Flags' (Item 2):

When-to-Read-Flags indicate, when to read or calculate a variable:

IRTFL   Root depth flag (1,2,3 possible)

1 = Root depth specified for the entire simulation.

2 = Root depth specified for each stress period.

3 = Root depth calculated as average for each time step from daily time series of root depth calculated from climate-data (T_{min}, T_{max}) read as time series for the entire simulation in Item 16 and a list of crop specific coefficients (coefficients for growing degree day calculation, polynomial coefficients, coefficients for root depth calculation) (Schmid and others, p. 47f) read for the entire simulation in Item 15.

ICUFL   Consumptive use flag (2,3 possible; **new FMP2-options for ICUFL = 1 or = –1**)

3 = FMP2 calculates a daily potential crop evapotranspiration flux ($ET_{c\text{-}pot}$) by multiplying a daily reference evapotranspiration flux (ET_{ref}) read as time series for the entire simulation in Item 16 with a daily crop coefficient K_c derived from parameters read for the entire simulation as Item 15 ($ET_{c\text{-}pot} = K_c * ET_{ref}$). FMP2 multiplies a daily $ET_{c\text{-}pot}$ averaged over each time step by the area of each cropped cell (ICID(IC,IR > 0) to yield a cell-by-cell $ET_{c\text{-}pot}$ flow rate for each time step. FMP2 multiplies the daily ET_{ref} flux averaged over each time step by the area of each fallow cell (ICID(IC,IR) = –1) to yield a cell-by-cell ET_{ref} flow rate for each time step. The ET_{ref} is assumed to be 100% evaporative for fallow cells where no transpiration exists.

2 = A list of crop specific fluxes of potential crop evapotranspiration ($ET_{c\text{-}pot}$) is read as Item 27a (Crop-ID, $ET_{c\text{-}pot}$ flux) for every stress period. FMP2 multiplies this ETc-pot flux by the area of the each cropped cell (ICID(IC,IR) > 0) to yield a cell-by-cell $ET_{c\text{-}pot}$ flow rate for each stress period. FMP2's fallow-cell option (ICID(IC,IR) = –1) cannot be used because no ET_{ref} flux is read if ICUFL = 2.

1 = A list of crop specific fluxes of potential crop evapotranspiration ($ET_{c\text{-}pot}$) is read as Item 27a (Crop-ID, $ET_{c\text{-}pot}$ flux) for every stress period and a constant or 2D real array reference evapotranspiration ET_{ref} (NCOL,NROW) is read as Item 27b for every stress period. FMP2 multiplies the $ET_{c\text{-}pot}$ flux by the area of the cropped cell (ICID(IC,IR) > 0) to yield a cell-by-cell $ET_{c\text{-}pot}$ flow rate for each stress period. FMP2 multiplies the ET_{ref} flux by the area of each fallow cell (ICID(IC,IR) = –1) to yield a cell-by-cell ET_{ref} flow rate for each stress period. The ET_{ref} is assumed to be 100% evaporative for fallow cells where no transpiration exists.

–1 = A list of crop specific crop coefficients (K_c) is read as Item 27a (Crop-ID, K_c) for every stress period and a constant or 2D real array of reference evapotranspiration ET_{ref} (NCOL,NROW) is read as Item 27b for every stress period. FMP2 multiplies the K_c by the ET_{ref} flux and by the area of each cropped cell (ICID(IC,IR) > 0) to yield a cell-by-cell $ET_{c\text{-}pot}$ flow rate for each stress period. FMP2 multiplies the ET_{ref} flux by the area of each fallow cell (ICID(IC,IR) = –1) to yield a cell-by-cell ET_{ref} flow rate for each stress period. The ET_{ref} is assumed to be 100% evaporative for fallow cells where no transpiration exists.

IPFL   Precipitation flag (2,3 possible)

2 = Precipitation flux specified for the each stress period

3 = Precipitation flux calculated as average for each time step from daily time series of precipitation flux specified in climate-data time series read in Item 16 for the entire simulation.

IFTEFL   Fraction-of-transpiration-and-evaporation-of-crop-consumptive-use flag (1,2 possible)

1 = Transpiratory and evaporative fractions of consumptive use specified for the entire simulation.

2 = Transpiratory and evaporative fractions of consumptive use specified for each stress period.

IIESWFL   Fraction-of-inefficiency-losses-to-SW-runoff flag (0,1,2 possible)

0 = The fraction of inefficiency losses to surface-water runoff is proportional to the slope of ground surface. The slope is estimated by FMP by a third order finite difference method using all eight outer points of the 3 ×3 kernel surrounding the cell. At cells directly adjacent to the boundary or the corners of the grid domain, the slope is calculated by using a 2 × 3 or 2 × 2 kernel, respectively. There is no data input required for FIESWP and FIESWI if IIESWFL is zero.

1 = Fractions of in-efficient losses to surface-water runoff related to precipitation and irrigation specified for the entire simulation.

2 = Fractions of in-efficient losses to surface-water runoff related to precipitation and irrigation specified for each stress period.

IEFFL   Efficiency Flag (1, 2 possible)

1 = On-farm efficiency either as OFE(Farm-ID) per farm or as **OFE(Farm-ID, Crop-ID$_{NCROPS}$)** per farm and per crop specified for the entire simulation.

2 = On-farm efficiency either as OFE(Farm-ID) per farm or as **OFE(Farm-ID, Crop-ID$_{NCROPS}$)** per farm and per crop specified for each stress period.

Water Policy Flags (Item 2):

IEBFL   Efficiency Behavior Flag

For IEBFL = 0,1: Cell-by-cell efficiency does not vary with changing groundwater level, but cell-by-cell delivery may vary with changing groundwater level. However, farm efficiency may vary in response to reduced delivery during deficit irrigation (if IDEFFL = –1).

0 = Conservative Behavior – Cell-by-cell efficiency is held constant over time step with respect to changing groundwater level. Farm efficiency reset to specified efficiency at each stress period.

1 = Conservative Behavior – Cell -by-cell efficiency is held constant over time with respect to changing groundwater level. Farm efficiency reset to specified efficiency at each time step.

For IEBFL = 2,3: Cell-by-cell efficiency varies with changing groundwater level, but cell-by-cell delivery does not vary with changing groundwater level. However, farm delivery may vary in response to deficit irrigation (if IDEFFL = –1).

2 = Conservative Behavior – Cell-by-cell delivery is held constant over time step with respect to changing groundwater level (evaluation of initial total delivery requirement (TDR) per cell at first iteration of first time step of each stress period). Farm efficiency reset to specified efficiency at each stress period.

3 = Conservative Behavior – Cell-by-cell delivery is held constant over time step with respect to changing groundwater level (evaluation of initial total delivery requirement (TDR) per cell at first iteration of each time step). Farm efficiency reset to specified efficiency at each time step.

IROTFL   Crop rotation flag:

< 0 Crop Type changes temporally and spatially at every stress period (CID 2D array is read for each stress period)

= 0 No crop rotation (CID 2D array is read for the entire simulation)

> 0 No crop rotation (CID 2D array is read for the entire simulation), and IROTFL = Stress period that is equal to Non-Irrigation Season

IDEFFL   Deficiency Scenario flag:

–2 = Water Stacking

–1 = Deficit Irrigation

0 = "Zero Scenario" where no policy is applied and if demand exceeds supply, it is assumed to be supplied by other imported sources

1 = Acreage-Optimization

2 = Acreage-Optimization with Water Conservation Pool

(only if SFR is specified in Name File, if a diversion from a river segment into a diversion-segment is specified in the SFR input file, and if routed or semi-routed deliveries from such a diversion-segment into farms can occur (IRDFL = 1, -1 or ISRDFL = 1, 2).

IBEN   Crop-Benefits Flag (only to specify if IDEFFL > 0):

1 = crop benefits list read for the entire simulation

2 = crop benefits list read for each stress period

ICOST   Water-Cost Coefficients Flag (only to specify if IDEFFL > 0):

0 = lumped water cost coefficients for the entire simulation

1 = water cost coefficients for each farm for the entire simulation

2 = water cost coefficients for each farm for each stress period

Crop Consumptive-Use Flag (Item 2):

ICCFL

Concept used for the approximation of ET-fluxes with changing head:

1 • for consumptive use Concept 1 = plant-and soil-specific pseudo steady state transpiration approximated by analytical solution: A restriction of active root zone corresponding to anoxia- or wilting-related pressure heads is determined by FMP by using analytical solutions of a vertical pseudo steady state pressure head distribution over the depth of the total root zone. (FMP2 not linked to UZF1).

2 • for consumptive use Concept 2 = nonplant- and nonsoil-specific simplification of Concept 1. (FMP2 not linked to UZF1).

3 • for consumptive use Concept 1 = plant-and soil-specific pseudo steady state transpiration approximated by analytical solution: A restriction of active root zone corresponding to anoxia- or wilting-related pressure heads is determined by FMP by using analytical solutions of a vertical pseudo steady state pressure head distribution over the depth of the total root zone. (FMP2 linked to UZF1: FMP2 farm identification arrays linked to coinciding UZF1 infiltration arrays).

4 • for consumptive use Concept 2 = nonplant- and nonsoil-specific simplification of Concept 1. (FMP2 linked to UZF1: FMP2 farm identification arrays linked to coinciding UZF1 infiltration arrays).

Surface-Water Flags (Item 2):

INRDFL

Non-Routed Surface-Water Delivery Flag:

0 = no Non-Routed Surface-Water Delivery exists.

1 = Non-Routed Surface-Water Deliveries exist. A farm related list of Volumes, Ranks, and Use-Flags of Non-Routed Surface-Water Delivery will be read.

MXNRDT

Maximum number of non-routed delivery types (read if INRDFL = 1).

ISRDFL

Semi-Routed Surface-Water Delivery Flag:

0 = no Semi-Routed Surface-Water Delivery exists

1 or 2 = Semi-Routed Surface-Water Deliveries exist. (Routing surface-water along a river or major canal, and allocating non-routed deliveries from a point of diversion). A farm related list of Row- and Column coordinates for a point of diversion will be read (only if SFR1 is specified in Name File).

1 = List of Row- and Column-coordinates is read for the entire simulation.

2 = List of Row- and Column-coordinates is read for each stress period.

IRDFL

Routed Surface Water Delivery Flag:

0 = no surface-water delivery exists.

1 = fully-routed surface-water delivery may occur from the uppermost reach of a series of diversion-segment reaches located within a farm. Caution: Streamflow fully-routed through a conveyance network directly to a farm can only occur (1) if SFR is specified in Name File, (2) if at least one reach of a diversion segment is located within the farms, and (3) if streamflow is available.

−1 = fully-routed surface-water delivery may occur from the uppermost reach of a series of reaches of any type of stream segment located within a farm. Caution: Streamflow fully-routed through a conveyance network directly to a farm can only occur (1) if SFR is specified in Name File, (2) if at least one reach of a any type of segment is located within the farms, and (3) if streamflow is available.

ISRRFL

Semi-Routed Surface-Water Runoff Returnflow Flag:

0 = No locations along the stream network are specified for any farm, where semi-routed runoff returnflow is recharged into the stream network. Runoff is either automatically prorated over non-diversion-segment reaches located within a farm, or automatically recharged into one non-diversion-segment reach nearest to the lowest elevation of the farm.

1 or 2 = For each farm, a location is specified anywhere along the stream network, where semi-routed runoff returnflow is recharged anywhere in the active model domain. A farm-related list of row and column coordinates or segment and reach numbers for a point of runoff returnflow recharge will be read (only if SFR is specified in Name File).

1 = List of row and column coordinates or segment and reach numbers is read for the entire simulation.

2 = List of row and column coordinates or segment and reach numbers is read for each stress period.

IRRFL   Routed Surface-Water Runoff Returnflow Flag:

0 = no surface-water runoff returnflow possible (no SFR specified in Name File).

1 = surface-water runoff may be returned as fully-routed returnflow to a series of non-diversion-segment reaches located within a farm (prorated over each reach weighted by the reach length). Caution: Fully-routed returnflow directly from a farm to a series of non-diversion-segment reaches can only occur (1) if SFR is specified in Name File and (2) if at least one reach of a non-diversion segment is located within the farm.

–1 = fully surface-water runoff may be returned as fully-routed returnflow to a series of reaches of any segment of stream type located within a farm (prorated over each reach weighted by the reach length). Caution: Fully-routed returnflow directly from a farm to a series of reaches of any type of stream segment can only occur (1) if SFR is specified in Name File and (2) if at least one reach of a non-diversion segment is located within the farm.

IALLOT   Surface-water allotment flag

0–No surface-water allotment specified,

1–Equally appropriated surface-water allotment height [L] specified per stress period (specification of diversions from a river into diversion segments in SFR input file required if ISRDFL = 1 or 2, or IRDFL = 1).

2–Prior appropriation system with Water Rights Calls [L^3/T] (diversion rates from a river into diversion segments are simulated if ISRDFL = 1 or 2, or IRDFL = 1; specification of a farm-specific water rights calls list required for each stress period).

3–Prior appropriation system without Water Rights Calls [L^3/T] (diversion rates from a river into diversion segments are simulated if ISRDFL = 1 or 2, or IRDFL = 1).

PCLOSE   User specified closure criterion for simulated diversions into diversion segments if prior appropriation is chosen [L^3/T] (only to specify if IALLOT > 1)

Mandatory Print Flags (Item 2):

IFWLCB   Farm well budget print flags

       < 0   A list (farm-well ID, farm ID, layer, row, column, farm-well flow rate) is printed to list file for time steps, for which in Output Control "Save Budget" is specified (using words) or ICBCFL is not (using numeric codes)

       = 0   farm-well flow rates not written

       = 1   A list (farm-well ID, farm ID, layer, row, column, farm-well flow rate) is saved on ASCII file "FWELLS.OUT" for all time steps

       > 1   if "Compact Budget" is not specified in Output Control:

                 A cell-by-cell 2D-array of farm-well flow rates will be saved as binary file on a unit number specified in the Name File for time steps, for which in Output Control "Save Budget" is specified (using words) or ICBCFL is not zero (using numeric codes).

           if "Compact Budget" is specified in Output Control:

                 A list (node, farm-well flow rate) will be saved as binary file on a unit number specified in the Name File for time steps, for which in Output Control "Save Budget" is specified (using words) or ICBCFL is not zero (using numeric codes).

IFNRCB   Farm net recharge budget print flags

       < 0   A cell-by-cell 2D-array of farm net recharge flow rates is printed to list file for time steps, for which in Output Control "Save Budget" is specified (using words) or ICBCFL is not zero (using numeric codes)

       = 0   farm net recharge flow rates not written

       = 1   A cell-by-cell 2D-array of farm net recharge flow rates is saved on ASCII file FNRCH_ARRAY OUT" for all time steps

       = 2   A list (stress period, time step, total time, farm ID, cumulative farm net recharge flow rates) will be saved as ASCII file "FNRCH_LIST.OUT"

       = 3   A list (stress period, time step, total time, farm ID, cumulative farm net recharge flow rates) will be saved as binary file "FNRCH_LIST_BIN.OUT" for all time steps

> 3	if "Compact Budget" is not specified in Output Control:
	A list (farm ID, cumulative farm net recharge flow rates) will be saved as binary file on a unit number specified in the Name File for time steps, for which in Output Control "Save Budget" is specified (using words) or ICBCFL is not zero (using numeric codes).
	if "Compact Budget" is specified in Output Control:
	if number of model layers = 1:
	A cell-by-cell 2D-array of farm net recharge flow rates will be saved as binary file on a unit number specified in the Name File for time steps, for which in Output Control "Save Budget" is specified (using words) or ICBCFL is not zero (using numeric codes).
	if number of model layers > 1:
	A 2D integer-array of each cells uppermost active layer, and a 2D real-array of each cell's farm net recharge flow rate will be saved as binary file on a unit number specified in the Name File for time steps, for which in Output Control "Save Budget" is specified (using words) or ICBCFL is not zero (using numeric codes).

ISDPFL   Farm supply and demand print flags

= –3	A list (A) of current demand and supply flow rates will be printed to the list file at each iteration, and a list (B) of final demand and supply flow rates will be printed to the list file for each time step: List (A): (FID, OFE, TFDR, NR-SWD, R-SWD, QREQ); List (B): (FID, OFE, TFDR, NR-SWD, R-SWD, QREQ, Q,[COMMENTS])
= –2	A list of final demand and supply flow rates will be printed to the list file for each time step: List: (FID, OFE, TFDR, NR-SWD, R-SWD, QREQ, Q, [COMMENTS])
= –1	A list of final demand and supply flow rates will be printed to the list file for time steps, for which in Output Control "Save Budget" is specified (using words) or ICBCFL is not zero (using numeric codes): List: (FID, OFE, TFDR, NR-SWD, R-SWD, QREQ, Q, [COMMENTS])
= 0	farm demand and supply flow rates not written
= 1	A list of initial demand and supply flow rates and of final demand & supply flow rates after the application of a deficiency scenario will be saved on ASCII file "FDS.OUT" for all time steps: List: (PER, TSTP, TIME, FID, OFE, TFDR-INI, NR-SWD-INI, R-SWD-INI, QREQ, TFDR-FIN, NR-SWD-FIN, R-SWD-FIN, QREQ, Q, DEF-FLAG)

> 1	if "Compact Budget" is not specified in Output Control:
	A list of initial demand & supply flow rates and of final demand and supply flow rates after the application of a deficiency scenario will be saved as binary file on a unit number specified in the Name File for all time steps
	List: list attributes are equal to ISDPFL = 1
	if "Compact Budget" is specified in Output Control:
	A list of initial demand & supply flow rates and of final demand & supply flow rates after the application of a deficiency scenario will be saved as binary file on a unit number specified in the Name File for time steps, for which in Output Control "Save Budget" is specified (using words) or ICBCFL is not zero (using numeric codes)
	List: list attributes are equal to ISDPFL = 1

IFBPFL   Farm budget print flags

= 0	Farm budget flow rates not written
= 1	A compact list of Farm Budget components (flow rates [L^3/T] and cumulative volumes [L^3] into and out of a farm) is saved on ASCII file "FB_COMPACT.OUT" for all time steps: List: (PER, TSTP, TIME, FID, Q-p-in, Q-sw-in, Q-gw-in, Q-ext-in, Q-tot-in, Q-et-out, Q-ineff-out, Q-sw-out, Q-gw-out, Q-tot-out, Q-in-out, Q-discrepancy[%], V-p-in, V-sw-in, V-gw-in, V-ext-in, V-tot-in, V-et-out, V-ineff-out, V-sw-out, V-gw-out, V-tot-out, V-in-out, V-discrepancy[%])

= 2　　A compact list of Farm Budget components (flow rates [L³/T] and cumulative volumes [L³] into and out of a farm) is saved on ASCII file "FB_COMPACT.OUT" for all time steps:

List: (PER, TSTP, TIME, FID,
Q-p-in, Q-nrd-in, Q-srd-in, Q-rd-in, Q-wells-in, Q-egw-in, Q-tgw-in, Q-ext-in, Q-tot-in,
Q-ep-out, Q-ei-out, Q-egw-out, Q-tp-out, Q-ti-out, Q-tgw-out, Q-run-out, Q-dp-out, Q-nrd-out,
Q-srd-out, Q-rd-out, Q-wells-out, Q-tot-out, Q-in-out, Q-discrepancy[%],
V-p-in, V-nrd-in, V-srd-in, V-rd-in, V-wells-in, V-egw-in, V-tgw-in, V-ext-in, V-tot-in,
V-ep-out, V-ei-out, V-egw-out, V-tp-out, V-ti-out, V-tgw-out, V-run-out, V-dp-out, V-nrd-out,
V-srd-out, V-rd-out, V-wells-out, V-tot-out, V-in-out, V-discrepancy[%])

> 2　　if "Compact Budget" is not specified in Output Control:

A list of farm budget flow rates will be saved as binary file on a unit number specified in the Name File for all time steps

List: list attributes are equal to IFBPFL =1 if unit number >2 is odd or equal to IFBPFL = 2 if unit number > 2 is even.

if "Compact Budget" is specified in Output Control:

A list of farm budget flow rates will be saved as binary file on a unit number specified in the Name File for time steps, for which in Output Control "Save Budget" is specified (using words) or ICBCFL is not zero (using numeric codes)

List: list attributes are equal to IFBPFL = 1 if unit number >2 is odd or equal to IFBPFL = 2 if unit number > 2 is even.

Optional Print Flags (Item 2):

IRTPFL　　Optional routing information print flag if the SFR Package is specified in Name file.
Information regarding the routing of farm deliveries and farm runoff returnflows will be written either to the listing file or to a separate ASCII file, called ROUT.OUT.

The information regarding deliveries tells whether the farm can potentially receive either:
(a) fully-routed deliveries from the first, most upstream located reach of a sequence of automatically detected delivery-segment reaches within a farm, or whether
(b) the farm can potentially receive semi-routed deliveries from specified stream reaches.

The information regarding returnflows tells whether potential runoff from the farm is returned either
(a) full-routed to automatically detected returnflow-segment reaches within a farm, over which the runoff returnflow is prorated, weighted by the length of each reach, or
(b) semi-routed to specified stream reaches, or in lack of this first two options,
(c) semi-routed to automatically detected returnflow-segment reach nearest to the lowest elevation of a farm.

= −2　　Routing information written to the listing file for the first stress period only.
= −1　　Routing information written to the listing file for every stress period.
= 0　　Routing information not written.
= 1　　Routing information written to ASCII file "ROUT.OUT" for every stress period.
= 2　　Routing information written to ASCII file "ROUT.OUT" for the first stress period only.

Options IRTPFL = −2 or 2 may be chosen if the geometry and the diversion rules specified in the SFR Package do not change from stress period to stress period.

IOPFL　　Optional print settings if Acreage-Optimization is chosen (if IDEFFL > 0)
= −4　　A tableaux matrix will be printed to the list file for iterations, during which optimization occurs.
= −3　　Original and optimized flow rates of resource constraints and a list of fractions of optimized cell areas will be printed to the list file for any farm and iteration that are subject to optimization:

List:	(Row,	Column,	Crop ID,	A-tot-opt/ A-tot-opt,	A-gw-opt/ A-tot-opt,	A-sw-opt/ A-tot-opt,	A-nr-opt/ A-tot-opt)

= –2   Original and optimized flow rates of resource constraints will be printed to the list file for any farm and iteration that are subject to optimization

= –1   A cell-by-cell 2D-array of fractions of active cell acreage will be printed to the list file for all time steps.

= 0   No original & optimized flow rates, and no optimized cell areas are written.

= 1   A cell-by-cell 2D-array of fractions of active cell acreage is saved on ASCII file "ACR_OPT.OUT" for all time steps.

= 2   Original and optimized flow rates of resource constraints are saved on ASCII file "ACR_OPT.OUT" for any farm and iteration that are subject to optimization.

= 3   Original and optimized flow rates of resource constraints and a list of fractions of optimized cell areas is saved on ASCII file "ACR_OPT.OUT" for any farm and iteration that are subject to optimization:

List:	(Row,	Column,	Crop ID,	A-tot-opt/ A-tot-opt,	A-gw-opt/ A-tot-opt,	A-sw-opt/ A-tot-opt,	A-nr-0pt/ A-tot-opt)

= –4   A tableaux matrix is saved on ASCII file "ACR_OPT.OUT" for iterations, during which optimization occurs.

IPAPFL   Optional print settings if Prior Appropriation is chosen (if IALLOT > 1)

= –1   A budget at the point of diversions from the river into diversion segments and a budget at the point of a farm-diversion from the diversion segment will be printed to the list file for all iterations.

= 1   A budget at the point of diversions from the river into diversion segments and a budget at the point of a farm-diversion from the diversion segment will be saved on ASCII file "PRIOR.OUT" for all iterations.

Flags for Auxiliary Variables (Item 2):

NOAUX   Indicates that no optional flags for auxiliary variables are specified. NOAUX is only required if Flag Blocks are used. If flags are read in Item 2b from a single line (as before in FMP1), then no entry is required if no optional flags for auxiliary variables are specified.

AUX "abc"   Defines an auxiliary variable, "abc", which will be read for each farm-well as part of Items 4 and 22. Up to five auxiliary attributes "abc" can optionally be specified, each of which must be preceded by "AUX." These values will be read after the QMAXfact or QMAX variable of Item 4 or Item 22, respectively.

AUX QMAXRESET   The specification of the optional flag "AUX QMAXRESET" for {option} in Item 2 will prompt FMP to reset QMAX as simulated by the MNW Package to the default QMAX as defined by FMP at the beginning of each time step. The optional flag "AUX QMAXRESET requires FMP to read an auxiliary variable after the QMAXfact or QMAX variable of the farm wells list in Items 4 or 22, or after any other preceding auxiliary variable (e.g., AUX NOCIRNOQ). If a "1" is read, then the MNW-simulated QMAX is reset to the default QMAX in the first iteration of each time step.

AUX NOCIRNOQ   The specification of the optional flag "AUX NOCIRNOQ" for {option} in Item 2 will prompt FMP to limit the distribution of farm pumpage to farm wells, whose row and column coincides with a top layer cell with a current irrigation requirement from active crops. "NOCIRNOQ" stands for "no crop irrigation requirement (CIR), no pumping (Q)." The optional flag "AUX NOCIRNOQ" requires FMP to read an auxiliary variable after the QMAXfact or QMAX variable of the farm wells list in Item 4 or 22, or after any other preceding auxiliary variable (e.g. AUX QMAXRESET). The auxiliary variable for "AUX NOCIRNOQ" is defined to be a binary parameter that tells which wells are selected for the NOCIRNOQ option. If a "1" is read, then the respective well is selected for setting its maximum capacity to zero if, during a particular time step, no crop irrigation requirement of the top layer cell exists. At each new time step the maximum capacity of such a select well will be reset to the default value.

Flags for Options (Item 2):

NOOPT   Indicates that no Options are specified. NOOPT is only required if Flag Blocks are used. If flags are read in Item 2b from a single line (as before in FMP1), then no entry is required if no Options are specified.

CBC   Indicates that memory should be allocated to store cell-by-cell flow for each well to make these flows available for use in other process.

NOPRINT   Indicates that a list of specified farm well attributes will not be printed to the list file.

WELLFIELD   Allows a series of irrigated farms to receive their cumulative irrigation demand as simulated non-routed deliveries from well fields simulated as virtual farms. A virtual well-field farm with one or several wells (well fields) receives a cumulative pumping requirement equal to the cumulative irrigation delivery requirement of irrigated farms that are supplied by the well field. If the cumulative demand exceeds the cumulative maximum pumping capacity of the well field, then other well field can supply the residual demand. The cumulative pumpage of the well field that is equal or less than the desired demand will then be re-distributed to the farms supplied by the well field weighted by the total delivery requirement (or residual delivery requirement for lower priority well fields) of the receiving farms. FMP2 then applies this re-distributed rate as non-routed deliveries to the respective farms.

For farms that receive water from a particular well field, in Item 33, the non-routed delivery volume may be set to a dummy zero, as the non-routed delivery is simulated by the well-field option. The rank of the non-routed delivery, NRDR, must consistently be equal to the priority of the well-field. The NRDU flag has to be set to "minus the Farm ID of the virtual farm that contains the well field" for the farms receiving water from the respective well field. For the virtual well-field farm itself, the NRDU flag has to be set to one.

For first priority well field and farms receiving water from that well field:

$\mathrm{NRDV}t_1(\mathrm{NFARMS})$ = 0 (dummy zero: simulated when option WELLFIELD is set)
$\mathrm{NRDR}t_1(\mathrm{NFARMS})$ = 1 (Type 1 must be of rank 1 for well-field farm and for receiving farms)
$\mathrm{NRDU}t_1(\mathrm{FID}_{rec\text{-}wf\text{-}1})$ = negative value of Farm-ID of virtual well-field farm
$\mathrm{NRDU}t_1(\mathrm{FID}_{wf\text{-}1})$ = 1
$\mathrm{NRDU}t_1(\mathrm{FID}_{other})$ = 0

For second priority well field and farms receiving water from that well field:

$\mathrm{NRDV}t_2(\mathrm{NFARMS})$ = 0 (dummy zero: simulated when option WELLFIELD is set)
$\mathrm{NRDR}t2(\mathrm{NFARMS})$ = 2 (Type 2 must be of rank 2 for well-field farm and for receiving farms)
$\mathrm{NRDU}t_2(\mathrm{FID}_{rec\text{-}wf\text{-}2})$ = negative value of Farm-ID of virtual well-field farm
$\mathrm{NRDU}t_2(\mathrm{FID}_{wf\text{-}2})$ = 1
$\mathrm{NRDU}t_1(\mathrm{FID}_{other})$ = 0

For a well field of priority *n* and farms receiving water from that well field:

$\mathrm{NRDV}t_n(\mathrm{NFARMS})$ = 0 (dummy zero: simulated when option WELLFIELD is set)
$\mathrm{NRDR}t_n(\mathrm{NFARMS})$ = *n* (Type *n* must be of rank *n* for well-field farm and for receiving farms)
$\mathrm{NRDU}t_n(\mathrm{FID}_{rec\text{-}wf\text{-}n})$ = negative value of Farm-ID of virtual well-field farm
$\mathrm{NRDU}t_n(\mathrm{FID}_{wf\text{-}n})$ = 1
$\mathrm{NRDU}t_n(\mathrm{FID}_{other})$ = 0
NRDV, NRDR, NRDU   definitions see "Non-Routed Surface-Water Deliveries" below
$\mathrm{FID}_{rec\text{-}wf\text{-}n}$ = Farm-ID of a farm receiving water from well-field *n*;
$\mathrm{FID}_{wf\text{-}n}$ = Farm-ID of a virtual well-field farm *n*.

The non-routed delivery type that originates from the lowest priority well field cannot be higher than the maximum number of non-routed delivery types, MXNRDT.

RECOMP_Q_BD   Re-computation of the Farm Process FM-routine is invoked at the end each time step loop.

MNWCLOSE   Head- and residual-closure criteria of the MODFLOW solver Package will be adjusted to allow convergence of the FMP pumping requirement to pumping simulated by the linked MNW Package.

QCLOSE   Criterion for actual MNW pumping rate to converge to FMP pumping requirement (real number)

HPCT   Fraction of reduction of head-change closure criterion if QCLOSE was not met [ ].

RPCT   Fraction of reduction of residual-change closure criterion if QCLOSE was not met [ ].

(QCLOSE, HPCT, and RPCT are optional and are only read if the MNWCLOSE option is specified.

Farm-Well Related Variables (Items 3, 4, 21, 22, 23)

Farm Well Parameter Definition (Item 3):

PARNAM	Parameter name for list of parameter farm-wells (called for each stress period to activate a list of parameter wells). This name can consist of 1 to 10 characters and is not case sensitive.
PARTYP	Parameter type (the only allowed parameter type is QMAX, which defines values of the volumetric maximum well capacity).
PARVAL	Parameter value (multiplier applied to parameter-wells).
NLST	Number of parameter farm-wells included in the parameter-well-list related to one parameter.
INSTANCES	Optional keyword that designates a parameter as time varying. The keyword is not case sensitive; that is, any combination of the same characters with different case can be used. If INSTANCES is present, it must be followed by a value for NUMINST. If INSTANCES is absent, PARNAM is non-time varying and NUMINST should not be present.
NUMINST	Number of instances for a time-varying parameter, where each instance is a list of wells and associated maximum capacities. If the keyword INSTANCES is present, it must be followed by a value for NUMINST. If INSTANCES is absent, NUMINST should not be present.
INSTNAM	Name of an instance associated with the parameter named in the corresponding Item 3. The instance name can be 1 to 10 characters and is not case sensitive. That is, any combination of the same characters with different case will be equivalent. Instance names must be unique for a parameter, but instance names may be reused for different parameters.

Farm-Wells List (Items 4, 22)

Layer	Layer number of cell containing the farm-well (for farm-wells linked to multi-layer wells defined in the Multi-Node Well Package: Layer No. = 0)
Row	Row number of cell containing the farm well
Column	Column number of cell containing the farm well
Farm-Well-ID	Farm-well identity associated with the farm well (to establish a link of a farm-well to a well defined in the Multi-Node Well Package: use "negative" Farm-Well-ID)
Farm-ID	Farm identity to which the farm-well is attributed
QMAXfact	Maximum Well Capacity factor (QMAXfact × PARVAL = QMAX) [L^3/T].
QMAX	Maximum Well Capacity [L^3/T] (positive value = maximum possible discharge rate)
[xyz]	Represents any auxiliary variables for a farm-well that have been defined in Item 2. The auxiliary variables must be present in each repetition of Items 4 and 22. If the optional flag for {Option} in Item 2 is set to "AUX QMAXRESET," then the auxiliary variable for [xyz] in column 7 of the farm wells list is defined to be a binary parameter that tells when to reset the MNW-simulated QMAX rate to the FMP-defined default QMAX rate. The parameter in column 7 of the well list is ignored, if the option flag "AUX QMAXRESET" is not specified. 0 = The MNW-simulated QMAX is reset at the beginning of each stress period. 1 = The MNW-simulated QMAX is reset at the beginning of each time step.

Farm-Well Flags and Dimensions for each Stress Period (Items 21, 23):

ITMP	Flag and counter > 0 ITMP = number of nonparameter farm-wells read for the current stress period. = 0 no nonparameter farm-wells are read for the current stress period. < 0 nonparameter farm-well data from the last stress period will be reused.

NP   Number of multiplier parameters in use in the current stress period.

Pname   Name of multiplier parameter being used in the current stress period. NP parameter names will be read.

Iname   Instance name read only if Pname is a time-varying parameter. Multiple instances of the same parameter are not allowed in a stress period.

Two-Dimensional Arrays (Items 5, 6, 8, 10 or 25, 27a, 30)

GSURF (NCOL, NROW)   Ground-surface elevation (Item 5)
FID (NCOL, NROW)   Farm identity (Item 6)
SID (NCOL, NROW)   Soil type identity (Item 8)
CID (NCOL, NROW)   Crop type identity (Item 10 if IROTFL $\geq$ 1, Item 25 if IROTFL = –1)
ETR(NCOL, NROW)   Reference Evapotranspiration (see climate related data) [L/T] (new Item if ICUFL= 1,–1)
PFLX (NCOL, NROW)   Precipitation flux (see climate related data) [L/T] (Item 30)

Farm Related Data Lists (Items 7, 19, 20a, 20b, 24, 32, 33, 34a, 34b, 36)

Farm-ID   Farm identity to which the parameters below are attributed
OFE   On-farm Efficiency per farm (real number between 0 and 1; 0. < OFE $\leq$ 1.), or
OFE(Farm-ID, Crop-ID$_{NCROPS}$)   On-farm Efficiency per farm and per crop (real number; 0. < OFE $\leq$ 1.),
GWcost1,2,3,4 and SWcost1,2,3,4   definitions see "Water Cost Coefficients" below
NRDV, NRDR, and NRDU   definitions see "Non-Routed Surface-Water Deliveries" below
Row, Column, Segment, Reach   definitions see "Locations of Diversion for Semi-Routed Surface-Water Deliveries" or "Locations of Returnflow for Semi-Routed Surface-Water Runoff" below
CALL   definitions see "Surface-Water Allotment\Prior Appropriation"

Soil Type Related Data List (Item 9)

Soil-ID   Soil type identity to which the parameters below are attributed
CapFringe   Capillary Fringe [L]

The following parameters are only needed if ICCFL = 1:

Either:
A-Coeff, B-Coeff, C-Coeff   Coefficients a, b, c for function DRZ = f($T_{c\text{-}pot}$, TRZ)
D-Coeff, E-Coeff   Coefficients d, e for function n = f(DRZ)

Or:
Soil-Type   Soil type in words:
                        3 options are available:
                        SANDYLOAM, SILT, and SILTYCLAY (not case-sensitive).
                        (For these three options, the FMP code contains intrinsic soil type specific coefficients a,b,c and d,e for the functionalities DRZ = f($T_{c\text{-}pot}$, TRZ) and n = f(DRZ). If a soil type is entered as a word, then a,b,c,d,e are not read).

The intrinsic coefficients in the program are as follows (Schmid, 2004):

Soil type	a	b	c	d	e
SANDYLOAM	0.201	−0.195	3.083	3.201	−3.903
SILT	0.320	−0.329	2.852	1.303	−2.042
SILTYCLAY	0.348	−0.327	1.731	0.530	−0.377

The parameters DRZ and n allow the fitting of a vertical pseudo steady state pressure head distribution over the total root zone:

(1)   The Depleted Root Zone (DRZ) is a function of the potential Transpiration and the Total Root Zone. It is defined as the lower part of the root zone at which the pressure head increases with depth from the minimum (negative) pressure head (defined as ψ4 in stress response function, see below) to zero at the bottom of the root zone.

$$DRZ = [exp(a \cdot \ln(TRZ \cdot MLT)b \cdot \ln(TPOT \cdot MLT) + c)]$$

(2)   The Sinuosity Coefficient (n) expresses the curvature of the vertical pressure head configuration over depth, which increases with increasing DRZ.

$$NEXP = d \cdot \ln(DRZ) + e$$

Although the intrinsic parameters a,b,c,d,e were derived based on CENTIMETER length units, multipliers in the program (MLT) can adjust the equations accordingly to length units of METER or FEET, if so chosen as LENUNI = 2 or = 1 in the Discretization file (Harbaugh and others, 2000).

Crop-Type-Related Data List (Natural Crop Growth Parameters) (Items 11–15, 25–29)

Crop-ID	Crop type identity, to which the parameters below are attributed
ROOT	Depth of root zone [L]
CU	Crop consumptive use flux [L/T] if ICUFL = 1, 2; **crop coefficient if ICUFL = –1**
NONIRR	Non-irrigation flag:
	0 = crop type is irrigated (zero value can be omitted)
	1 = crop type is not irrigated
FTR	Transpiratory fraction of consumptive use ($0 < FTR \leq 1$)
FEP	Evaporative fraction of consumptive use related to precipitation ($0 \leq FEP \leq 1$)
FEI	Evaporative fraction of consumptive use related to irrigation ($0 \leq FEI < 1$)
FIESWP	Fraction of in-efficient losses to surface-water related to precipitation ($0 \leq FIESWP \leq 1$)
FIESWI	Fraction of in-efficient losses to surface-water related to irrigation ($0 \leq FIESWI \leq 1$)
PSI(1)	Negative or positive value of pressure head, at which root uptake becomes zero due to anoxia or high pressure [L]
PSI(2)	Negative or positive values of pressure head, at which root uptake is at maximum and from which uptake decreases with rising pressure head due to anoxia [L]
PSI(3)	Negative pressure head, at which root uptake is at maximum and from which uptake decreases with falling pressure head due to wilting [L]
PSI(4)	Negative pressure head, at which root uptake becomes zero due to wilting [L]
BaseT	Base temperature
MinCutT	Minimum cutoff temperature
MaxCutT	Maximum cutoff temperature
C_0, C_1, C_2, C_3	Polynomial coefficients for CGDD – K_c functionality (see Chapter "General Data Requirements")
BegRootD	Beginning root depth [L]
MaxRootD	Maximum root depth [L]
RootGC	Root growth coefficient

Climate-Related Data (Items 16, 27b, 30)

Climate Time Series (Item 16):

TimeSeriesStep	Time-step in climate time series. The length of a time series time step must consistently be equal to the MODFLOW time unit chosen in the Discretization File (ITMUNI).
	For ICUFL = 3 or IRTFL = 3, the MODFLOW time unit must be days (ITMUNI = 4). For IPFL = 3 (while ICUFL ≠3, and IRTFL ≠3), all MODFLOW time units are possible (seconds, minutes, hours, days, years; ITMUNI = 1, 2, 3, 4, 5). However, ITMUNI = 1 or 2 for units of seconds or minutes should be avoided for very long periods of simulation due to the possibility of insufficient computer memory.
Precip	Precipitation flux [L/T]
MaxT	Maximum temperature
MinT	Minimum temperature
ETref	Reference Evapotranspiration flux [L/T]
LENSIM	Total number of steps in a time series = length of the simulation.

Reference Evapotranspiration Array (Item 27b):

ETR(NCOL,NROW)   Reference Evapotranspiration Array or Constant [L/T] if ICUFL = 1, −1

Precipitation Array (Item 30):

PFLX(NCOL, NROW)   Precipitation Flux Array or Constant [L/T]

Crop-Type-Related Data Lists (Agro-Economic Parameters) (Items 17, 18, 31)

Fallow List (Item 17):

Crop-ID   Crop-type identity to which the parameter below is attributed.
IFallow   Fallow-Flag:
               1 = Crop type fallowed
               0 = Crop type not fallowed (for example, pecan trees)

Crop Benefits List (Items 18, 31):

Crop-ID   Crop-type identity to which the parameters below are attributed.
WPF-Slope   Slope of crop-specific water-production function (yield vs. $ET_{c\text{-}act}$)
WPF-Int   Intercept of crop-specific water-production function (yield vs. $ET_{c\text{-}act}$) (can be zero).
Crop-Price   Market-price per crop [value/weight]

Water Cost Coefficients (Items 19, 32)

Farm-ID   Farm identity to which the cost coefficients below are attributed.

Groundwater Cost Coefficients:

GWcost1   Groundwater Base Maintenance Costs per unit volume [$/L^3$]
GWcost2   Groundwater Costs for Pumping in Well per unit volume, per unit lift [$/(L^3 \cdot L)$]
GWcost3   Groundwater Costs for Vertical Lift from Well to Cell per unit volume, per unit lift [$/(L^3 \cdot L)$]
GWcost4   Groundwater Delivery Costs per unit volume, per unit distance [$/(L^3 \cdot L)$]

Surface-Water Cost Coefficients:

SWcost1   Fixed Price of (Semi-) Routed Surface-Water per unit volume [$/L^3$]
SWcost2   Costs for Vertical Lift of (Semi-)Routed Surface-Water from Reach to Cell per unit volume, per unit lift [$/(L^3 \cdot L)$]
SWcost3   Delivery Costs of (Semi-)Routed Surface-Water per unit volume, per unit distance [$/(L^3 \cdot L)$]
SWcost4   Fixed Price of Non-Routed Surface-Water per unit volume [$/L^3$]

Non-Routed Surface-Water Deliveries – Farm-Related Data List (Item 33)

Farm-ID   Farm identity to which the parameter below are attributed
NRDV   Volume of Non-Routed Delivery Type [L^3]
NRDR   Rank of Non-Routed Delivery Type
NRDU   Binary "NRD_{use}-Flag" of Non-Routed Delivery Type:
               if 0: only the amount sufficient is used to meet the farm's demand,
               if 1: the absolute amount available is used; surplus discharged back into farm's head gate reach.
               if 2: the absolute amount available is used; surplus injected into farm wells attributed to farm ID.

Locations of Diversion for Semi-Routed Surface-Water Deliveries (Items 20a, 34a) or Locations of Returnflow for Semi-Routed Surface-Water Runoff (Items 20b, 34b) – Farm-Related Data Lists

Farm-ID   Farm identity to which the parameter below are attributed

Row   Row number of point of diversion (for ISRDFL > 0) or returnflow (for ISRRFL > 0)

Column   Column number of point of diversion (for ISRDFL > 0) or returnflow (for ISRRFL > 0)

Segment   Number of stream segment in which the diversion reach (for ISRDFL > 0) or returnflow reach (for ISRRFL > 0) is located (must be equal to the number of the identical stream reach specified in column four of the data list defined in the SFR2 input file defined for the entire simulation)

Reach   Number of reach from which diversion (for ISRDFL > 0) or to which the returnflow (for ISRRFL > 0) occurs (must be equal to the number of the identical reach specified in column five of the data list in the SFR2 input file defined for the entire simulation)

Four options of data input (marked by "x") are available in order to uniquely identify the point of diversion or returnflow within a cell:

Row	Column	Segment	Reach	Comments	
x	x	x	x	Full set of information is available	Maximum information
x	x	x	0/–	If more than one segment passes through the cell	User prefers identification of location by row/column coordinates
x	x	0/–	0/–	If just one segment passes through the cell	
0	0	x	x	If more than one segment pass through the cell	User prefers identification of location by segment and reach number

Surface-Water Allotment (Items 35, 36)

Equal Appropriation:

ALLOT   Surface-water allotment height [L] for a stress period.

Prior Appropriation:

Farm-ID   Farm identity to which the parameter below is attributed

CALL   Water Rights Call attributed to a farm [L^3/T]

Output Data for Farm Process

Simulation results from FMP2 can be reported to seven auxiliary data sets in addition to the main MF2005 listing and global files. These data sets consist of the following components. Various options to either print these data to the MF2005 listing file or to save them to ASCII or binary files are controlled by the associated flags in parentheses:

(1)   Farm-well budget (IFWLCB);

(2)   Farm net-recharge budget (IFNRCB);

(3)   Farm supply and demand budget (ISDPFL);

(4)   Farm Budget – Budget of all physical flows into and out of a farm (IFBPFL)

(5)   Routing information for farm deliveries and returnflows (IRTPFL);

(6)   Optimized flow rates and optimized acreage of farms that experience a deficiency (IOPFL); only if acreage optimization is chosen as a deficiency scenario (IDEFFL > 0);

(7)   Budget at the point of diversions from the river into diversion segments and a budget at the point of a farm diversion from the diversion segment (IPAPFL); only if prior appropriation is chosen as surface-water rights option (IALLOT > 1).

Farm-Well Budget

The simulated farm-well flow rates can either be printed for each well location identified by layer, row, and column to the list file or saved to an ASCII file named "FWELLS.OUT." In addition, farm-well flow rates can be saved to a binary file for each well location identified by the respective model node, or as a 2D-array for each cell.

Farm Net-Recharge Budget

Simulated farm net recharge flow rates can be printed as a 2D-array for each cell to the list file or to an ASCII file named "FNRCH_ARRAY.OUT." Alternatively, a list of the cumulative farm net recharge for each farm can be saved to either an ASCII file named "FNRCH_LIST.OUT" or to a binary file named "FNRCH_LIST_BIN.OUT." In addition, a list of cumulative farm net-recharge flow rates or a 2D-array of cell-by-cell farm net recharge flow rates can optionally be saved to binary files on a unit number specified in the Name File.

Farm Supply and Demand Budget

The simulated components of farm irrigation demand and supply of any current stage of solution during a time step (per iteration) and of the final demand and supply rates at the end of time steps or stress periods may be printed to list file:
Lists of current (iterative) and final farm demand and supply flow rates consist of the following parameters:
(1) FID farm ID
(2) OFE on-farm efficiency
(3) TFDR total farm delivery requirement
(4) NR-SWD non-routed surface-water delivery
(5) R-SWD (semi-)routed surface-water delivery
(6) QREQ groundwater pumping requirement
(7) (Q-FIN) groundwater pumping only available for list of final rates)

Notice, that the list of "current" rates is iteratively updated within a present time step and does not yet contain a final supply flow rate from groundwater pumping, Q-FIN, which is available the end of a time step and therefore included into the list of final rates. For cases of irrigation water supply insufficiency, a comment is printed at the end of each record, informing about a possible imbalance of the farm demand and supply budget.

If the final supply exceeds the original or optimized demand by a certain flow rate X, then the following messages will be printed:

For Deficit Irrigation or Zero Scenario (IDEFFL = –1 or 0):
     "QREQ exceeds QMAXF by" X

For Deficit Irrigation with Water-Stacking (IDEFFL = –2):
     "Original QREQ exceeded QMAXF by" X
     "QREQ of priority crops still exceeds QMAXF by" X

If, for Acreage-Optimization (IDEFFL > 0), the optimized demand is actually less than the original constrained surface-water or groundwater resource (by a flow rate of X), then the following messages will be printed:
     "Surface-Water Demand falls behind original Surface-Water Supply by" X
     "Groundwater Demand falls behind original QMAXF by" X

Another, more comprehensive form of saving initial and final farm demand and supply budget is to save the according flow rates to an ASCII file named "FDS.OUT" for all time steps or alternatively to a binary file either for all or for selected time steps. Final rates only differ from initial ones, if either water-stacking or acreage-optimization was applied as deficiency scenario.

A list of initial and final farm demand and supply flow rates for all time steps consists of the following parameters:

General Information:
(1) PER:	Stress period
(2) STP:	Time step
(3) TIME [UNIT]:	Elapsed time (in units chosen in discretization file)
(4) FID:	Farm identification
(5) OFE:	Specified or calculated on-farm efficiency

Initial flow rates before invoking a deficiency scenario:

(1) TFDR-INI:	Initial Total Farm Delivery Requirement
(2) NR-SWD-INI:	Initial Non-Routed Surface-Water Delivery
(3) R-SWD-INI:	Initial (Semi-) Routed Surface-Water Delivery
(4) QREQ-INI:	Initial Pumping Requirement

Final flow rates of a solution found by means a deficiency scenario:

(1) TFDR-FIN:	Final Total Farm Delivery Requirement
(2) NR-SWD-FIN:	Final Non-Routed Surface-Water Delivery
(3) R-SWD-FIN:	Final (Semi-) Routed Surface-Water Delivery
(4) QREQ-FIN:	Final Pumping Requirement
(5) Q-FIN:	Final Pumping Rate
(6) DEF-FLAG:	Deficiency Scenario Flag

Farm Budget

A list of flow rates, Q [L^3/T], or cumulative volumes, V [L^3], of the simulated Compact Farm Budget components is saved for all time steps in ASCII file "FB_COMPACT.OUT" (if IFBPFL = 1) or in a binary file on a unit number specified in the Name File (if IFBPFL > 2 and odd). The list is saved in a binary file for all time steps if "Compact Budget" is not specified in Output Control or for time steps, for which in Output Control "Save Budget" is specified, if "Compact Budget" is specified in Output Control.

A list of Compact Farm Budget rates consists of the following parameters:

Headers in Farm Budget	Explanation
Model attributes:	
PER	Stress period
STP	Time step
TIME	Time unit chosen in discretization file (example "DAYS" if ITMUNI = 4 in MF Discretization File)
FID	Farm ID
Flow rates into a farm:	
Q-p-in	Precipitation
Q-sw-in	Surface water inflow
Q-gw-in	Groundwater inflow
Q-ext-in	External deliveries
Q-tot-in	Total inflows
Flow rates out of a farm:	
Q-et-out	Evapotranspiration outflow
Q-ineff-out	Inefficient losses
Q-sw-out	Surface water outflow (excess non-routed deliveries back into stream segment)
Q-gw-out	Groundwater outflow (excess non-routed deliveries injected into farm-wells)
Q-tot-out	Total outflows
Q-in-out	Inflows minus Outflows
Q-Discrepancy[%]	Percent discrepancy

A list of flow rates, Q [L^3/T], or cumulative volumes, V [L^3], of the simulated Detailed Farm Budget components is saved for all time steps in ASCII file "FB_DETAILS.OUT" (if IFBPFL = 2) or in a binary file on a unit number specified in the Name File (if IFBPFL > 2 and even). The list is saved in a binary file for all time steps if "Compact Budget" is not specified in Output Control or for time steps, for which in Output Control "Save Budget" is specified, if "Compact Budget" is specified in Output Control.

A lists of Detailed Farm Budget rates consist of the following parameters:

Headers in Farm Budget	Explanation
Model attributes:	
PER	Stress period
STP	Time step
TIME	Time unit chosen in discretization file (example "DAYS" if ITMUNI = 4 in MF Discretization File)
FID	Farm ID
Flow rates into a farm:	
Q-p-in	Precipitation
Q-nrd-in	Non-routed deliveries
Q-srd-in	Semi-routed deliveries
Q-rd-in	Fully routed deliveries
Q-wells-in	Groundwater well pumping deliveries
Q-egw-in	Evaporation from groundwater into the farm
Q-tgw-in	Transpiration from groundwater into the farm
Q-ext-in	External deliveries
Q-tot-in	Total inflows
Flow rates out of a farm:	
Q-ei-out	Evaporation from irrigation out of the farm
Q-ep-out	Evaporation from precipitation out of the farm
Q-egw-out	Evaporation from groundwater out of the farm
Q-ti-out	Transpiration from irrigation out of the farm
Q-tp-out	Transpiration from precipitation out of the farm
Q-tgw-out	Transpiration from groundwater out of the farm
Q-run-out	Overland runoff out of the farm
Q-dp-out	Deep percolation out of the farm
Q-nrd-out	Non-routed deliveries from the farm
Q-srd-out	Semi-routed deliveries out of the farm (in form of excess non-routed deliveries recharged back into remote' head-gate)
Q-rd-out	Fully routed deliveries out of the farm (in form of excess non-routed deliveries recharged back into a head-gate within the farm)
Q-wells-out	Injection from farm into farm-wells (excess non-routed deliveries injected into farm-wells)
Q-tot-out	Total outflows
Q-in-out	Inflows minus outflows
Q-Discrepancy[%]	Percent discrepancy

For both the compact and the detailed farm budget, cumulative volumes [L^3] into and out of a farms are printed to the right of the flow rates and are denoted by "V" analogous to "Q" for flow rates (e.g.: V-p-in = cumulative precipitation into a farm).

Routing Information for Farm Deliveries and Runoff Returnflows

The following illustrates the format, in which the routing information for a particular farm is written either to the listing file or to file ROUT.OUT. Depending on how the user has set flags IRDFL, ISRDFL, IRRFL, and ISRRFL, one statement out of several possible statements (separated by OR) will inform about the routing system of deliveries or runoff returnflows. Text highlighted in yellow is text that is written to either the listing file or to file ROUT.OUT. Exactly which information is written is explained in text highlighted in light blue:

```
ROUTING INFORMATION FOR FARM:           ?
---------------------------------------
DELIVERIES:

  FULLY-ROUTED DELIVERIES:
```

```
Information is given on whether the search for reaches of diversion segments (IRDFL=1) or of any type of
segments(IRDFL=-1) within a farm is activated or deactivated.
If IRDFL=0, or if IRDFL=1 or =-1 and ISRDFL>0 and a point of diversion for semi-routed delivery has been
specified already anywhere along the stream network, then this search is deactivated.
Previously, in FMP1, the user was notified by an error message if an automatically found reach within
a farm was indeed available as diversion head-gate and, at the same time, a location for a semi-routed
delivery was specified. InFMP2, any user-specified location of a stream reach for a semi-routed delivery
(when ISRDFL>0) takes precedence over an automatically available reach within a farm for a fully-routed
delivery (when IRDFL=1 or =-1). Once a userspecified stream reach for a semi-routed delivery is detected,
the code skips the search for reaches within a farm.
```

```
ACTIVATED SEARCH FOR REACHES OF DIVERSION SEGMENTS THAT ARE WITHIN A FARM
OR
ACTIVATED SEARCH FOR REACHES OF ANY STREAM SEGMENTS THAT ARE WITHIN A FARM
OR
DEACTIVATED SEARCH FOR REACHES OF DIVERSION SEGMENTS THAT ARE WITHIN A FARM
OR
DEACTIVATED SEARCH FOR REACHES OF ANY STREAM SEGMENTS THAT ARE WITHIN A FARM
OR
ROUTED DELIVERY OPTION WAS NOT SELECTED
```

```
Information is given on the locations of reaches found automatically within a farm, and of the reach,
which is the first, most upstream reach used as head-gate for full-routed diversions to a farm. If no
reaches were found within a farm, then information is given that a full-routed diversion is not possible.

FULLY ROUTED DELIVERY FROM THE FIRST, MOST UPSTREAM REACH OF A SEQUENCE OF REACHES
THAT ARE WITHIN THE FARM:
  HEAD-GATE WITHIN THE FARM AT:
   ROW  COLUMN  SEGMENT NO.  REACH NO.
    ?      ?         ?           ?
  SEQUENCE OF REACHES THAT ARE WITHIN THE FARM:
   ROW  COLUMN  SEGMENT NO.  REACH NO.
    ?      ?         ?           ?
    ?      ?         ?           ?
ACTIVE FARM DELIVERY-SEGMENT LENGTH   ????.????
OR
NO ACTIVE FARM DELIVERY-SEGMENT REACHES ARE WITHIN THE FARM: NO FULLY-ROUTED DIVERSION POSSIBLE
```

SEMI-ROUTED DELIVERIES:

```
Information is given on the location of a stream reach specified for a diversion of a semi-routed
delivery. If ISRDFL=0 or if ISRDFL>0 and no reach was specified for a particular farm, then information
is given that a semi-routed diversion is not possible.

SEMI-ROUTED DELIVERY FROM A SPECIFIED STREAM REACH AT:
    ROW   COLUMN  SEGMENT NO.   REACH NO.
     ?       ?         ?            ?
OR
NO POINT OF DIVERSION FOR SEMI-ROUTED DELIVERY SPECIFIED: NO SEMI-ROUTED DIVERSION POSSIBLE
```

RETURNFLOWS:

FULLY-ROUTED RETURNFLOWS:

```
Information is given on whether the search for reaches of non-diversion segments (IRRFL=1) or of any
type of segments (IRRFL=-1) within a farm is activated or deactivated. Unless ISRDFL>0 and a point of
semi-routed runoff returnflow has been specified anywhere on the stream network, this search is always
activated as FMP attempts to return runoff fully-routed to reaches within a farm. This is attempt based
on the assumption, that occurring runoff always has to be returned to the stream network if possible in
order to preserve mass. Therefore, the user does not have the option to disable the check for reaches
receiving fully-routed returnflow analogous to a check for reaches, which fully-routed deliveries are
diverted from.
If ISRRFL>0 and a point of semi-routed runoff-returnflow has been specified anywhere on the stream
network, then this search is deactivated.
Previously, in FMP1, the user was notified by an error message if automatically found reaches within a
farm were indeed available to receive fully-routed runoff-returnflow and, at the same time, locations for
semi-routed runoff-returnflows were specified. In FMP2, any user-specified location of a stream reach for
a semi-routed runoff-returnflow (when ISRRFL>0) takes precedence over automatically available reaches
within a farm for fully-routed returnflows. Once a user-specified stream reach for a semi-routed delivery
is detected, the code skips the search for delivery-segment reaches adjacent or within a farm.

ACTIVATED SEARCH FOR REACHES OF NON-DIVERSION SEGMENTS THAT WITHIN A FARM

OR

ACTIVATED SEARCH FOR REACHES OF ANY STREAM SEGMENTS THAT WITHIN A FARM

OR

DEACTIVATED SEARCH FOR REACHES OF NON-DIVERSION SEGMENTS THAT ARE WITHIN A FARM

OR

DEACTIVATED SEARCH FOR REACHES OF ANY STREAM SEGMENTS THAT ARE WITHIN A FARM
```

```
Information is given on the locations of reaches found within a farm, over which fully-routed runoff-
returnflow from a farm is prorated, weighted by the length of each reach. If no reaches were found within
a farm, then information is given that a full-routed runoff returnflow is not possible.

FULLY ROUTED RUNOFF RETURNFLOW PRORATED OVER REACHES WITHIN THE FARM AT:
    ROW   COLUMN  SEGMENT NO.   REACH NO.
     ?       ?         ?            ?
     ?       ?         ?            ?
ACTIVE FARM RETURNFLOW-SEGMENT LENGTH   ????.????
OR
NO ACTIVE FARM RETURNFLOW-REACHES ARE ADJACENT OR WITHIN THE FARM: NO FULLY-ROUTED RETURNFLOW POSSIBLE
```

```
SEMI-ROUTED RETURNFLOWS:
```

```
Information is given on the location of a stream reach specified to receive semi-routed runoff
returnflow. If the primary search for reaches within a farm receiving fully-routed runoff returnflow is
negative, then a secondary search is executed by FMP for a reach of a non-diversion segment (IRRFL=1)
or of a segment of any type (IRRFL=-1) nearest to the lowest elevation of the farm's ground surface.
However, this secondary search is only executed if ISRRFL=0 or if ISRRFL>0 and no reach was specified for
a particular farm.
If neither an automatically found returnflow reach nor a specified stream reach is found that can receive
semi-routed runoff returnflow, then information is given that a semi-routed diversion is not possible.

SEMI-ROUTED RUNOFF RETURNFLOW TO A SPECIFED STREAM REACH AT:
     ?         ?          ?             ?
OR
SEMI-ROUTED RUNOFF RETURNFLOW TO A STREAM REACH FOUND NEAREST TO THE LOWEST FARM ELEVATION AT:
     ?         ?          ?             ?
OR
NO POINT OF RECHARGE FOR SEMI-ROUTED RETURNFLOW SPECIFIED: NO SEMI-ROUTED RETURNFLOW POSSIBLE
```

Optimized Flow Rates and Optimized Acreage of Farms

The user has various options of saving different data of interest, if acreage-optimization was chosen as a deficiency scenario ($IDEFFL > 0$). Fractions of active cell acreage will be printed as a 2D array either to the list file or saved to an ASCII file named "ACR_OPT.OUT" for all time steps. Alternatively, original and optimized flow rates of resource constraints may either be saved for each farm by themselves or in conjunction with a list of fractions of active cell acreage. This option will save the according data either to the list file or to an ASCII file named "ACR_OPT.OUT" for any farm and any iteration that are subject to optimization.

For each cell (row, column) within an optimized farm, the list of fractions of cell acreage consists of the following parameters:

A-tot-opt/A-tot-max	fraction of total optimized acreage on total maximum acreage;
A-gw-opt/A-tot-opt	fraction of optimized groundwater irrigated acreage on total optimized acreage;
A-sw-opt/A-tot-opt	fraction of optimized (semi-)routed surface-water irrigated acreage on total optimized acreage,
A-nr-opt/A-tot-opt	fraction of optimized non-routed surface-water irrigated acreage on total optimized acreage;

Users with specific interest in the definition of the linear optimization tableaux matrix may save this matrix either to the list file or to an ASCII file named "ACR_OPT.OUT." The number of columns in the tableaux matrix equals the number of optimization variables + 1. The number of rows in the matrix equals the number of constraints + 1.

Budgets at Points of Diversion from the River and Farm Diversion

A budget at the point of diversions from the river into diversion segments and a budget at the point of a farm-diversion from the diversion segment are printed to the list file or to an ASCII file named "PRIOR.OUT" if Prior Appropriation is chosen as surface-water allotment option ($IALLOT > 1$). The "Prior Appropriation Subroutine" in FMP solves (1) for the delivery to a farm from a diversion segment and (2) for the diversion into the respective diversion segment from a river. Solutions for (1) and (2) are found iteratively. The budgets for both points of diversion are therefore printed for any farm on an iterative basis. However, after having found solutions for (1) and for (2) for a certain farm, those solutions are not iterated anymore within a current time step. The output budgets for both points of diversion also informs whether a "convergence solution" or "exceedance solution" was found. A "convergence solution" is found if the surface-water delivery to the farm "convergences" to the delivery requirement from the farm's head-gate reach. An "exceedance solution" is found if the necessary diversion from the river into the respective diversion segment "exceeds" the river streamflow and consequently the diversion from the diversion segment into a junior water rights farm will be insufficient to satisfy the delivery requirement from that farm's head-gate reach.

The output data set for each farm consists of three blocks of information:

1. Information about routing system during current iteration:
   Farm-ID;
   Head-gate reach number;
   Delivery segment number;
   River segment number.

2. Budget at Point of Diversion from River into a Diversion Segment:

Qstr-in	Inflow to point of diversion at the end of current stream segment;
Qstr-out	Outflow from point of diversion past the end of current stream segment;
Qstr-min	Minimum river-streamflow requirement at point of diversion from stream that is not available for diversion to the current farm (necessary to account for the demand and for related conveyance losses to a downstream farm senior farm located at a downstream diverting segment);
ADIV	Actual diversion rate from stream into current delivery segment;
PDIV	Potential diversion rate from stream into current delivery segment.

3. Budget at Point of Farm Diversion from Diversion Segment:

RDEL-req	Delivery requirement from current head-gate reach;
Qcn-in	Inflow to point of diversion from current diversion segment at beginning of current head-gate reach;
Qcn-out	Outflow from point of diversion from current diversion segment past the beginning of current head gate reach;
Qcn-min	Minimum "canal-streamflow requirement" at point of diversion from diversion segment that is not available for farm "f" at its head-gate (necessary to account for senior farm on the same diversion segment);
DELIVERY	surface-water delivery to current farm from current head-gate reach at present iteration.

Comments:

STAGE:	RESULT:
A record (in quotes) is printed informing about the current stage of the "prior appropriation" algorithm; three different stages are possible.	A record (in quotes) is printed informing about the action taken at a\ certain stage of the algorithm; three different results are possible at two different stages
"INITIAL VALUES" (initial values at beginning of algorithm are printed)	
"CUMULATE PDIV" (values are printed after cumulating PDIV by the unsatisfied increment [RDELreq. – Qcn-in])	"Exit and apply new PDIV rates" (exit MF2005 and solve with incremented PDIV)
"SOLUTION" (final values are printed once a solution was found)	"Convergence Solution" "Exceedance Solution"

Example Problem

A hypothetical, but realistic, example problem is simulated with MF2005 using FMP2 jointly with the SFR2, UZF1, and MNW Packages to demonstrate the new linkages and flow interdependencies now available in MF2005-FMP2. Although all features of FMP2 are not included in this example, this example serves to illustrate many of the fundamental features needed for many regional hydrologic models that include simulations of supply and demand related to irrigated agriculture, natural vegetation and urban setting of water use and movement. The problem spans two calendar years with monthly stress periods. Selected input and output data sets are shown in Appendix A.

Model Structure and Input

The example problem is similar to the previous FMP1 example problem described by Schmid and others (2006). The example model shows a linkage between FMP2 and the streamflow routing network simulated by the SFR2 Package. The model used in the example problem includes four model layers to demonstrate the linkage between FMP2 and the MNW Package that simulates multi-node wells screened across several layers. To illustrate the new link between FMP2 and UZF1, two regions are included in the example model to demonstrate the simulation of delayed recharge beneath a farm through a deeper unsaturated zone and the simulation of rejected infiltration and groundwater discharge to the surface in a riparian area with high groundwater levels that slightly rise above the land surface. The farm, where the model simulates delayed recharge, is located in the northwestern part of the model domain where a deeper unsaturated zone exists between the land surface and the initial water table in the uppermost model layer (layer 1, fig. 6; fig. 8). The riparian area is located near the eastern model boundary adjacent to the outflow region of the through-flowing river (fig. 8).

The boundary conditions include stream-aquifer interaction simulated by the upgraded version of the Streamflow Routing Package (SRF2) and general head boundaries at the up-gradient (western) and down-gradient (eastern) edge of the model domain for layers 1 and 2 (fig. 7).

The example also includes temporally distributed precipitation that was typical of the rainfall for Davis, California in the Central Valley which helps to facilitate delayed recharge following more time-varying supplies from precipitation and irrigation to crops, urban areas, and native vegetation.

The example model uses the option that streambed elevations of diversion segments follow the slope of ground surface at a defined depth (Appendix B). The fully routed deliveries and returnflows of the previous FMP1 example model were replaced with semi-routed deliveries and returnflows (fig. 8) to illustrate how the user can control the routing system design.

The example model includes eight model farm IDs (fig. 8), six model crop-type IDs (fig. 9), and three model soil-type IDs (fig. 10). Even though in FMP2 all model cells do not necessarily need to be assigned to "model farm IDs," in this particular example, all model cells of the model domain were assigned to eight "virtual farms" that represent water-accounting regions. Six of these "virtual farms" are associated with farm wells for the potential delivery of groundwater if needed. (fig. 8). There are two additional non-irrigated, "rain-fed" water-accounting regions that represent a riparian wetland on the eastern boundary surrounding the river outflow (virtual farm 8) and natural vegetation throughout the remainder of the model (virtual farm 7).

The example model includes six virtual crop types that represent groups of crops aggregated by similar crop coefficients and growth-stage lengths (fig. 9). Although MF2005-FMP2 provides the option to change the spatial distribution of crop types from stress period to stress period (often called "crop rotation"), in the present example the distribution of crop types does not change over time. Crop type 1 represents vegetable row crops consisting of 20 percent cabbage, 50 percent lettuce, and 30 percent green beans. Crop type 2 represents apple, cherry, and walnut tree orchards. Crop type 3 represents for winter grains, such as barley, wheat, and oat. The landscaping of the urban area, crop type 4, represents lawns and gardens, which are simulated with crop coefficients of turf. Crop type 5 represents native vegetation comprising equal areas of pasture-grazed, grass-clover, wild-life area, and non-bearing trees and vines. Crop type 6 represents a riparian area with willows that are capable of taking up water under variable saturated conditions.

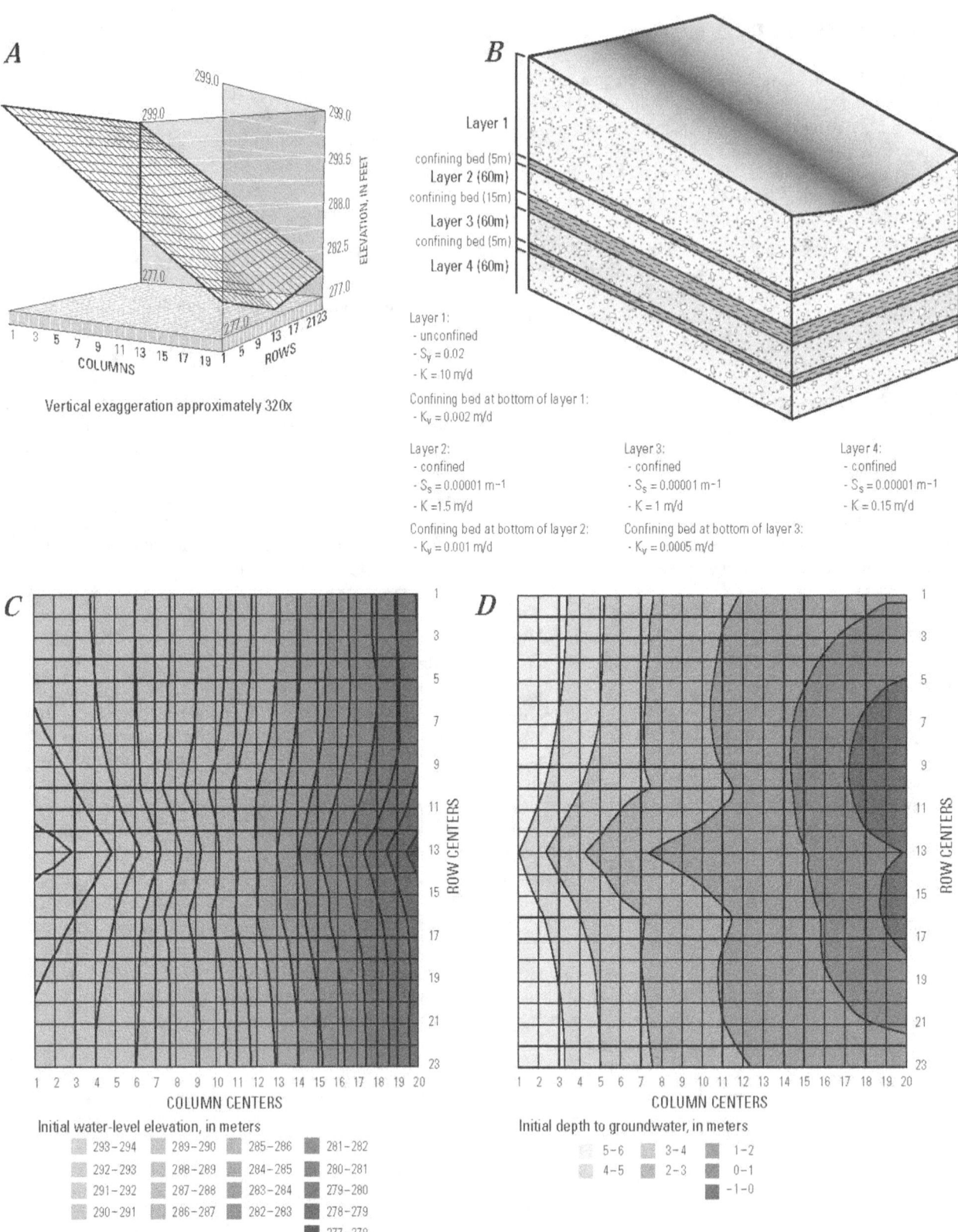

Figure 6.   (*A*) Land-surface elevation, (*B*) model layers, (*C*) nitial water levels in the uppermost model layer (layer1), and (*D*) the initial depth-to-water in model layer 1 for FMP2 example problem.

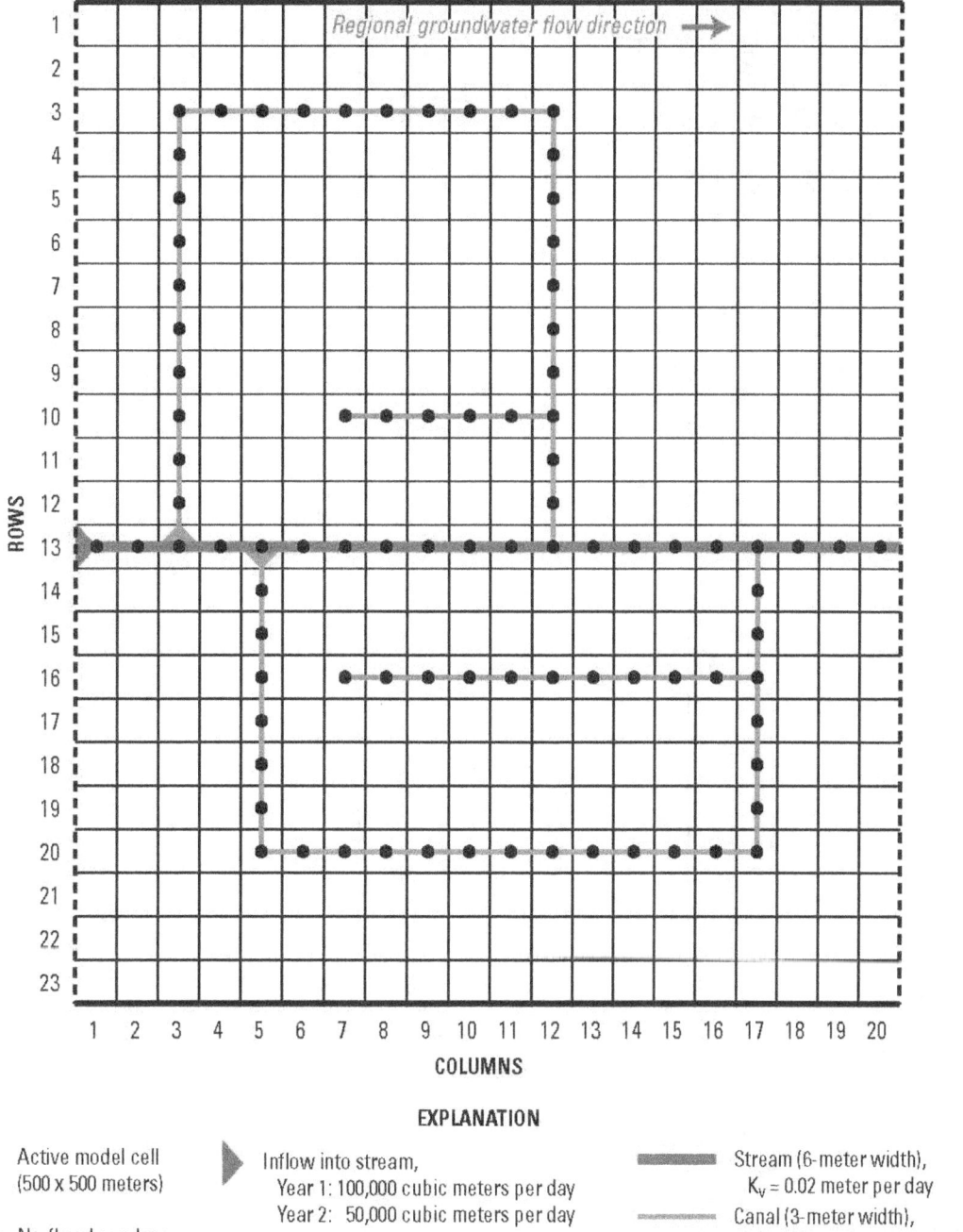

Figure 7. Model domain and boundary conditions for the FMP2 example problem.

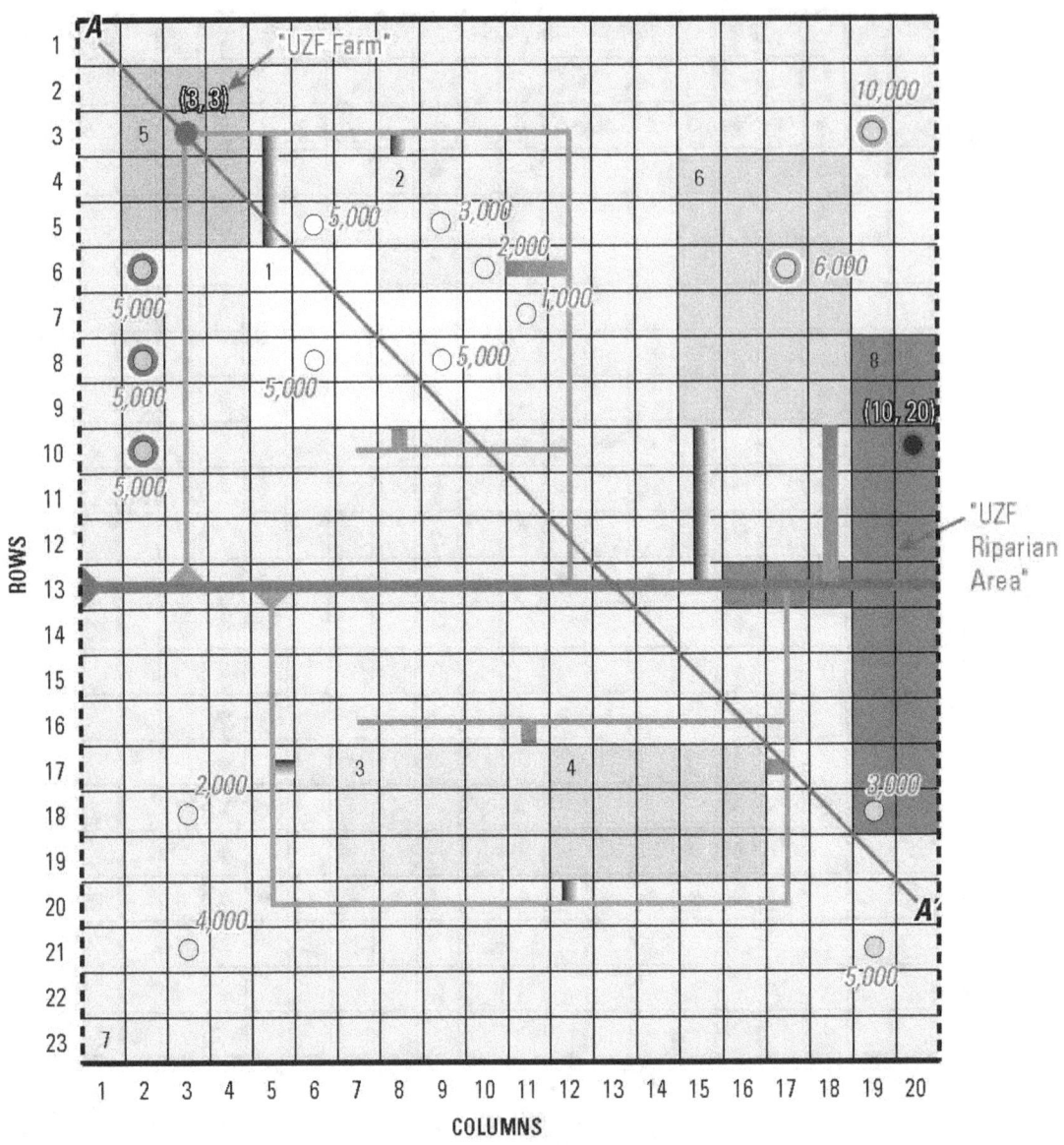

Figure 8. Model domain with farms and related wells for FMP2 example problem.

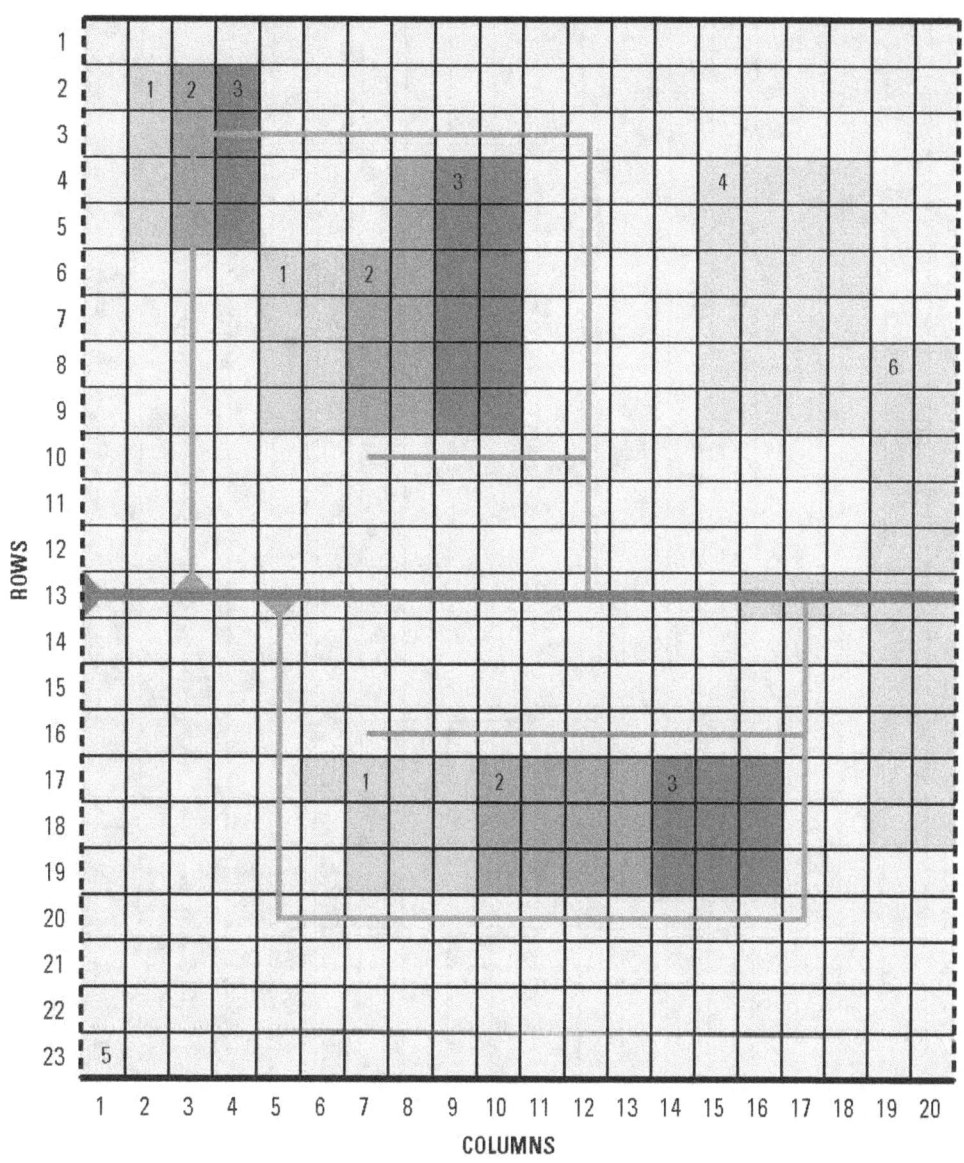

Figure 9. Map showing crop and other vegetation distribution for FMP2 example problem.

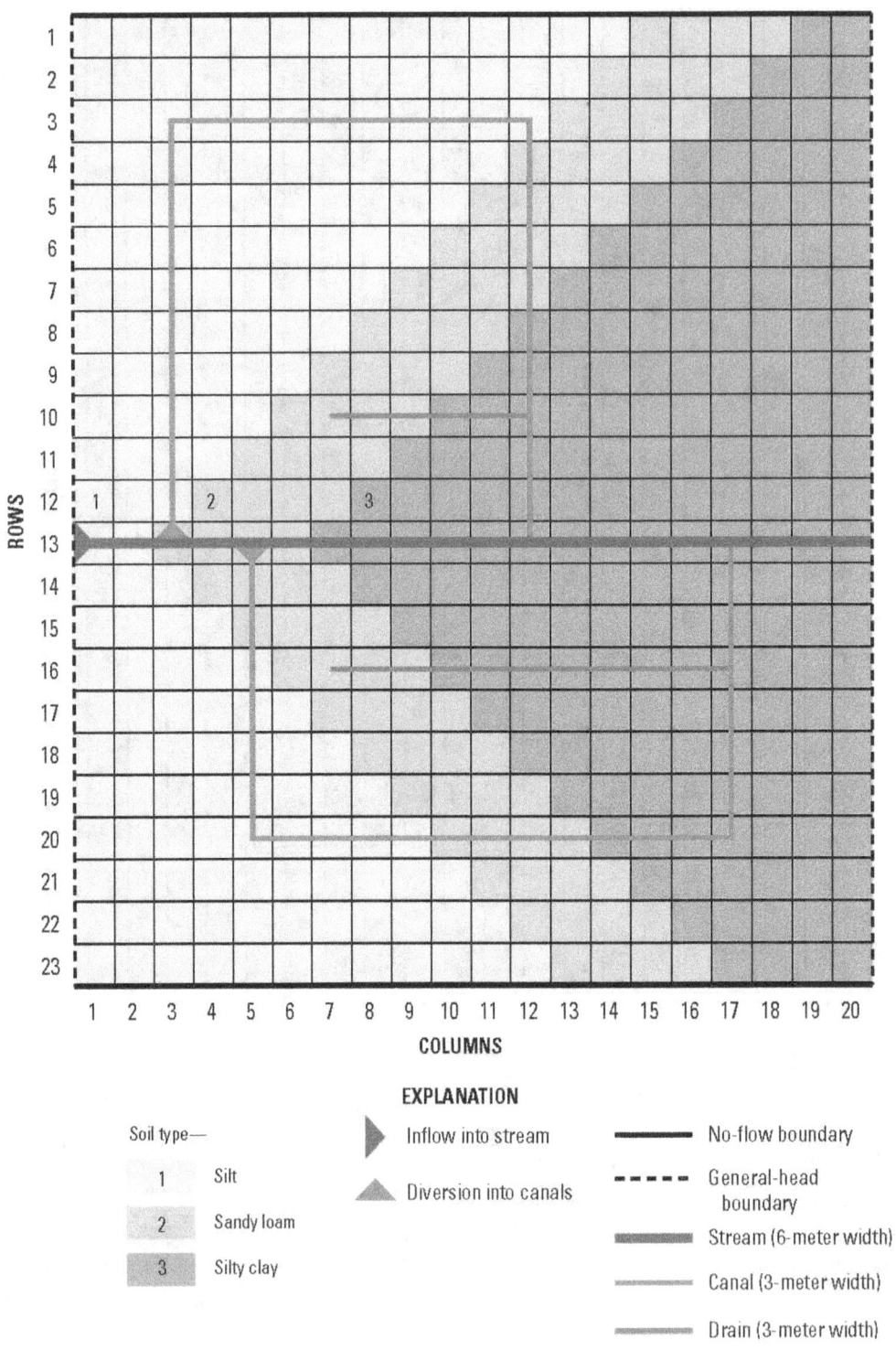

Figure 10.   Distribution of soils for FMP2 example problem.

For each crop group, weighted averages of individual crop coefficients and growth stage lengths were computed based on the percentage contribution of each individual crop. The individual values for initial-, mid-, and end-season crop coefficients as well as the durations for initial, development, mid, and late growth stages were compiled from databases of various sources of literature (for example, Food and Agriculture Organization Irrigation and Drainage Paper 56 in Allen and others, 1998). For each crop group that represents average growth and harvest attributes, a daily time series (365 days) of crop coefficients was calculated using the "composite" crop coefficients and "composite" growth stage lengths. Finally, twelve monthly averages of crop coefficients were calculated based on the daily time series and applied over the two-year simulation period to stress periods 1 through 12 and 13 through 24 of the example model. The monthly crop coefficients allow the different vegetation to be active at different times of the year as they each cycle through their seasonal growth stages (fig. 11A). Virtual crop coefficients for virtual crop types (crop groups) as described above were preprocessed for the example model prior to assembling the FMP2 data input. The technique and algorithms applied are formulated in EXCEL spreadsheets that also contain a compilation of literature crop coefficients and growth stage lengths. These EXCEL spreadsheets can be provided by the authors on request. Other approaches on how to preprocess crop coefficients for each model stress period may be possible.

Crop specific parameters in FMP2 include fractions of transpiration and fractions of evaporation related to precipitation or to irrigation for the six crop groups. The separate simulation of transpiration and evaporation is an essential difference of MF2005-FMP2 from many other hydrologic models, which assume a common extinction depth for a joint evapotranspiration term. In FMP2, the evaporation from groundwater is extinct at a depth to water equal to a specified capillary fringe, but the transpiration from groundwater is extinct at a depth to water equal to the root zone plus the capillary fringe. The example problem simulates crop transpiration under both unsaturated conditions (crop types 1 through 5) and saturated conditions (for example, crop type 6 simulated as riparian willows) by analytical solutions. Fractions of transpiration and evaporation are varied on a monthly basis (figs. 11B–D).

The fraction of transpiration, FTR, may be derived as $FTR = K_{cb}/K_c$ from the literature if in addition to the total crop coefficient, K_c, a "basal" transpiratory crop coefficient, K_{cb}, is also available (for example, Allen and others, 1998, Allen and others 2005). The fraction of evaporation that is related to exposed areas wetted by precipitation, FEP, depends on the exposed nonvegetative bare soil surface wetted by precipitation. Even though, in reality, transpiration and evaporation may be related nonlinearly, for the virtual crop types 1 through 3 and 5 in this example model, we simplify the fraction of evaporation to be equal to the complement of the fraction of transpiration, that is, FEP = 1 – FTR. The fraction of evaporation related to irrigation depends on the fraction of the exposed soil surface that is wetted by irrigation. Unlike soil surface wetted by precipitation, the exposed areas wetted by irrigation may not be entirely wetted. The extent to which the exposed area is wetted depends on the irrigation method used, which, in reality, often follows a particular crop type. For the virtual crop types 1 through 3 in the example model, the fraction of transpiration related to irrigation is assumed to be constrained by the lesser of the complement of the fraction of transpiration or by the wetted fraction, fw, that is available from the literature for certain irrigation methods (Allen and others, 2005, or Allen and others, 1998), that is, FEI = min(1 – FTR, fw). Fractions of transpiration and evaporation are FMP parameters that have a high uncertainty and MODFLOW models utilizing FMP are quite sensitive to these parameters (Schmid and others, 2008). The demonstrated approach is one of many ways that the fraction of transpiration and evaporation can be physically based or based on published data. Rough initial estimates of these fractions may be specified, but one is advised to improve these parameters with estimates gained during the model calibration process.

For the urban area (crop type 4), the fraction of transpiration is assumed to be equal to the fraction of the entire area from which transpiration takes place (for example, lawns and gardens). In many cases, land-use surveys specify the percentage of irrigated land in urban areas. In the example model, an average value of such a percentage range (for example, 12.5 percent as the average of 0 to 25 percent) is used to represent the fraction of the area (that is, 0.125), where transpiration occurs. The fraction of evaporation then is assumed to equal to the fraction of the entire urban area that is open and exposed (housing, parking lots, industry, airports).

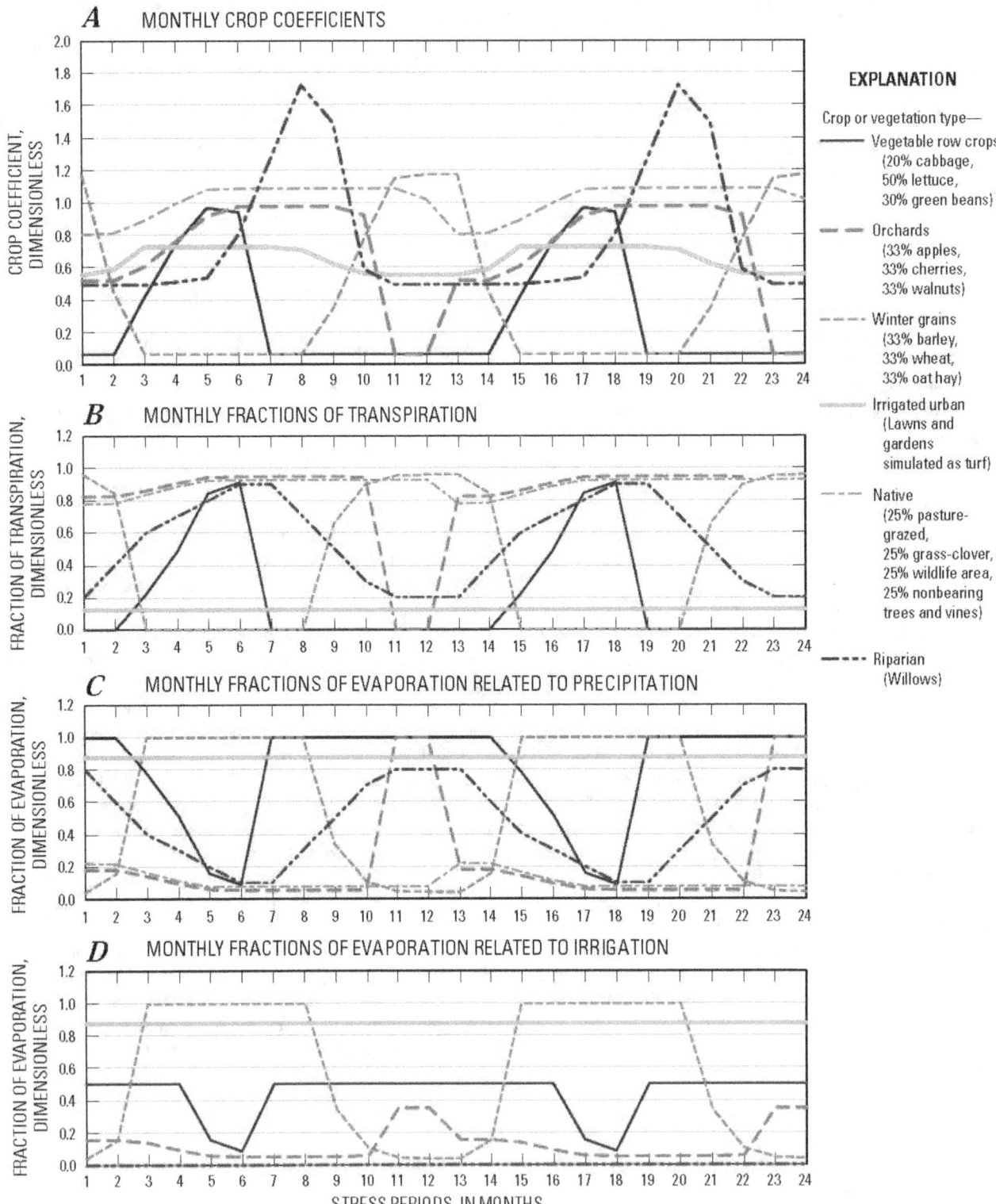

Figure 11. (A) Crop coefficients, (B) fractions of transpiration, (C) fractions of evaporation related to precipitation, and (D) fractions of evaporation related to irrigation through time for the six virtual crop types of the FMP2 example problem.

For the Native Vegetation (crop type 5), the fraction of evaporation related to irrigation is specified using placeholder zero values because no irrigation is applied. For Riparian Vegetation (crop type 6), the fractions of transpiration and of evaporation related to precipitation are pure assumptions. No basal crop coefficients, Kcb, were found in the literature that could be applied. The fraction of evaporation related to irrigation again are placeholder zero values because no irrigation is applied.

The model represents the three soil types that are internally defined by FMP as silt, sandy loam, and silty clay (fig. 10). Root depths are specified for all crop types for every stress period (IRTFL = 2) and vary for some of the crop types such as vegetable row crops and winter grains, while being held constant for the others. For the example model, the maximum rooting depth is taken to be the average between values available through Allen and others (1998, table 22) and Brush and others (2004). For perennial crops such as orchards and turf or for native and riparian vegetation, the rooting depth is assumed constant over time. For annuals like vegetable row crops and winter grains, the root zone depth is assumed to vary proportionally to the crop coefficient of each stress period with a proportionality factor equal to the ratio of maximum rooting depth to maximum crop coefficient. This algorithm is used as long as the crop coefficient increases or remains constant at its maximum or minimum.

$$RZ^t = (RZ_{max}/K_{c\text{-}max}) \times Kc^t, \quad \text{if} \quad K_c^t \geq K_c^{t-1} \quad \text{or} \quad K_c^t = K_{c\text{-}min}$$
$$RZ^t = RZt-1, \quad \text{if} \quad K_c^t < K_c^{t-1} \quad \text{or} \quad K_c^t \neq K_{c\text{-}min}$$

During the end period, the crop coefficient declines until harvest takes place. Yet, the root zone that reached a maximum during the mid period is assumed to remain at the maximum until the crop coefficient drops to the off-season minimum value corresponding to harvest or senescence.

Fractions of inefficient losses to surface-water runoff are specified for each virtual crop type for the each stress period. In MF2005-FMP2 surface-water runoff is assumed to depend on irrigation methods, which, in turn, may depend in part on the crop type. Because rainfall intensity and irrigation application methods further influence runoff, FMP2 requires input of two separate fractions of inefficient losses to surface-water runoff: one related to precipitation (FIESWP) and

another one related to irrigation (FIESWI), which may be omitted or set to placeholder zero values for non-irrigated crop types, such as native vegetation (crop type 5) and riparian (crop type 6). In the example model, FIESWP and FIESWI are held constant over time for crop types 1 through 4. However, FIESWP increases for native vegetation (crop type 5) and riparian (crop type 6) during the winter–spring months indicating an increased fraction of inefficient losses to runoff during the heavy winter–spring precipitation typical for the climate of Davis, California. Additional runoff components are calculated by the UZF1-FMP2 linkage for farm 5 and the riparian area (farm 8) stemming from infiltration in excess of the saturated hydraulic conductivity, the groundwater discharge to land surface, and rejected infiltration for high groundwater levels. In FMP2, two flags indicate the design of the runoff returnflow routing system (see below). In UZF1, a two-dimensional integer array, IRUNBD, specifies for each UZF-active cell the SFR streamflow segment where the potential runoff is returned to (Appendix A).

Crop-specific parameters, such as crop coefficients, root zone depths, fractions of transpiration and evaporation, and fractions of inefficient losses to surface-water runoff, can vary from stress period to stress period. Contrary to that, pressure heads that define stress-response function coefficients are the only crop-related set of parameters that are specified for the entire simulation. Noticeably, in FMP2, a stress response function can be defined under both unsaturated and saturated conditions for either negative or positive pressure heads, at which uptake is either zero or at maximum. In the example model simulation, the stress response of riparian willow trees (crop type 6) to water uptake is described by a stress response function, where the optimal uptake takes place under unsaturated conditions, but a reduced uptake is still possible in saturated conditions until the pressure head reaches 20 cm and uptakes becomes zero (Appendix A, file PSI.IN).

Reference evapotranspiration and precipitation are set constant over each monthly stress period but vary from stress period to stress period. The data are derived from CIMIS data for the weather station at the UC Davis in California (http://wwwcimis.water.ca.gov/cimis/data.jsp, accessed April 20, 2009). For each month of the year, a median was calculated from the monthly values during the period from 1982 to 2008.

Surface water deliveries to irrigated farms include non-routed water transfers from outside the model domain and equally appropriated semi-routed deliveries along a streamflow routing network simulated with the SFR2 Package. Non-routed deliveries (NRDs) are assumed to be known volumes of deliverable water for each stress period (Appendix A, file NRDV.IN). NRDs are supplied to all but the natural vegetation and riparian areas with a variable monthly scale factor that changes the volume of the NRDs over the course of each model year (Appendix A, file NRDFAC.IN). Semi-routed surface-water deliveries to irrigated farms are diverted from specified stream reaches (Appendix A, file SRD. IN) located outside the farm domain. The term "semi" is used in the sense of:

(a) deliveries routed along the stream network to a user-specified point of diversion;

(b) non-routed delivery (for example, pipe flow) from the user-specified point of diversion (perceived as 'remote head-gate') to the farm.

Semi-routed runoff is returned to the stream network (simulated by SFR2) at a specified location only for farm 1 (Appendix A, file SRR.IN). A new flag in FMP2, ISRRFL, was set to 1 indicating that these locations are specified only once for the entire simulation. For all farms other than farm 1, zeros are specified indicating that no returnflow location is specified and, if, alternatively, also no stream segment is located within the farm's domain, to automatically search for a stream reach located nearest to the lowest elevation of the farm, to which runoff returnflow will be discharged. The new flag in FMP2, IRRFL, was set to -1, so that the farm's runoff would be prorated as fully-routed returnflow over the reaches of "any type of stream segment" found within the farm (as opposed to "non-diversion segments" if IRRFL = 1). For three farms, farm 5, the native vegetation (farm 7), and the riparian area (farm 8), stream segments were found within the domain of each farm and each farm's returnflow was prorated accordingly over those reaches.

An output file ROUT.OUT was written that informs about the system of routing deliveries to, and runoff away from, each farm. Appendix A contains the portion of the file that pertains to stress period 1, time step 1.

The data input for linked packages can be found in Appendix A. The reader is referred to the SFR2, UZF1, and MNW input instructions regarding the explanations of the SFR2, UZF1, and MNW data input used in the example model (Niswonger and Prudic, 2005; Niswonger and others, 2006; Halford and Hanson, 2002). The streamflow network and its hydraulic properties are depicted in figure 7. The location and screening of multi-node wells is shown in figure 8.

The new linkage to the UZF1 Package facilitates delayed recharge through the unsaturated zone in the upgradient (western part of the model domain), such as at farm 5 (fig. 8). This linkage also allows simulation of rejected infiltration in the riparian areas (eastern part of the model domain), such as farm 8 (fig. 8). The areas where this linkage is active are specified through the UZF1 Package input in the IUZFBND array shown in figure 12. The additional unsaturated-zone properties that are specified include a Brooks–Corey epsilon of 0.35, a saturated water content of 0.2, an initial water content of 0.16, and a saturated vertical hydraulic conductivity of the unsaturated zone of 0.001 meters per day. The relation between the land surface and the initial water table and the peak-season water table for model layer 1 is shown for the unsaturated zone beneath farm 5 (fig. 13).

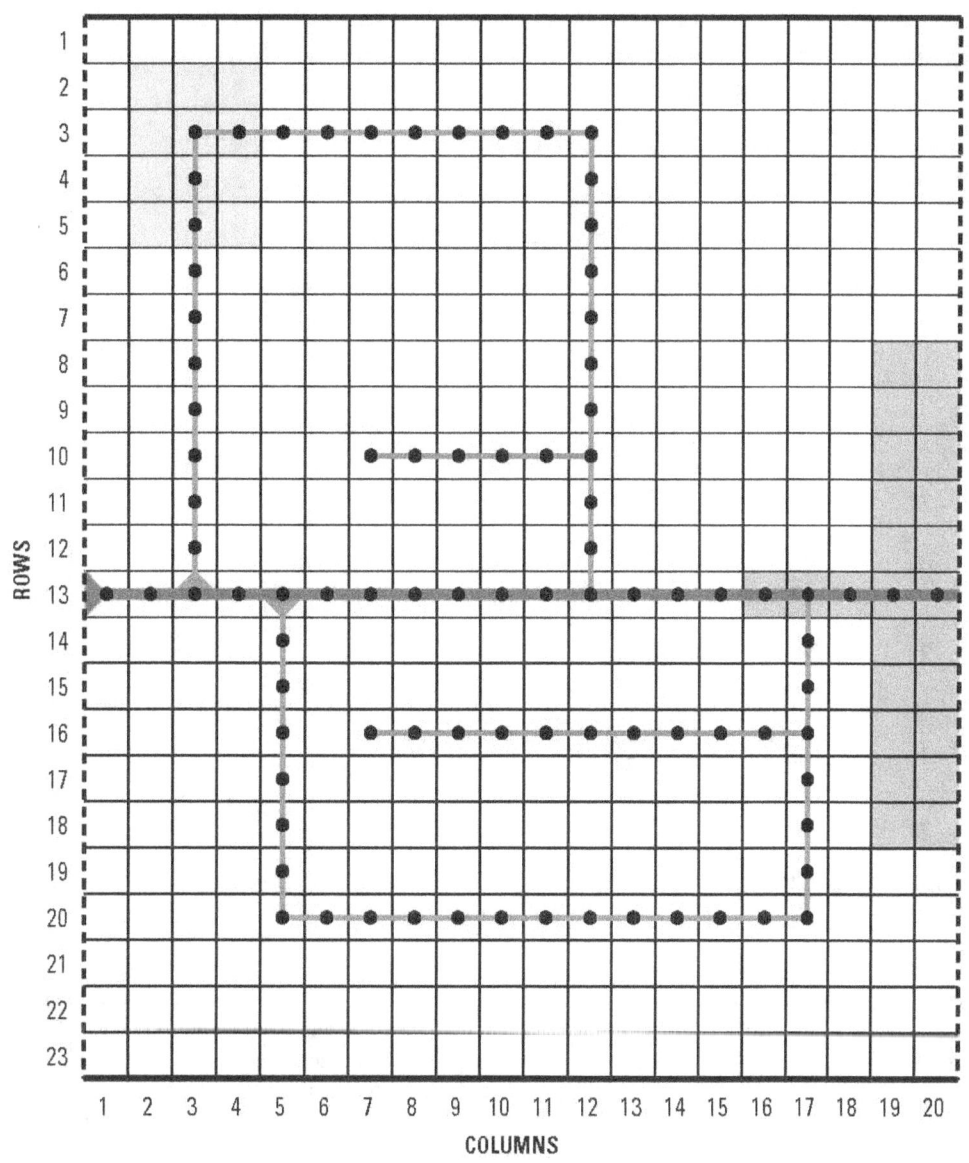

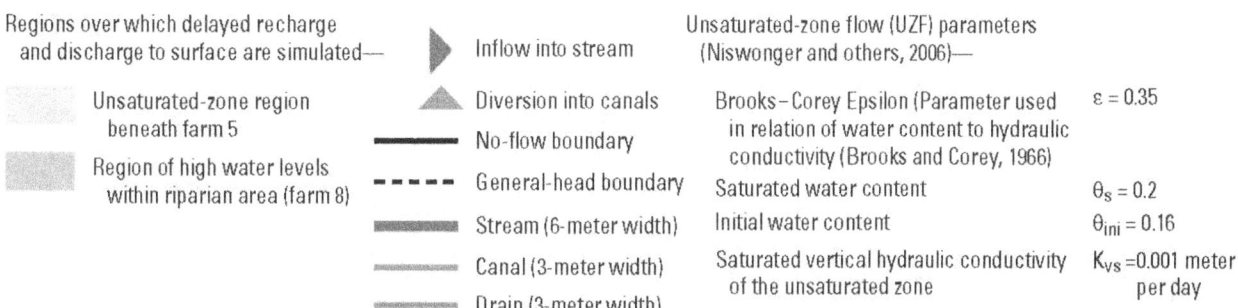

Figure 12. Distribution of zones linked to UZF Package for unsaturated-zone flow.

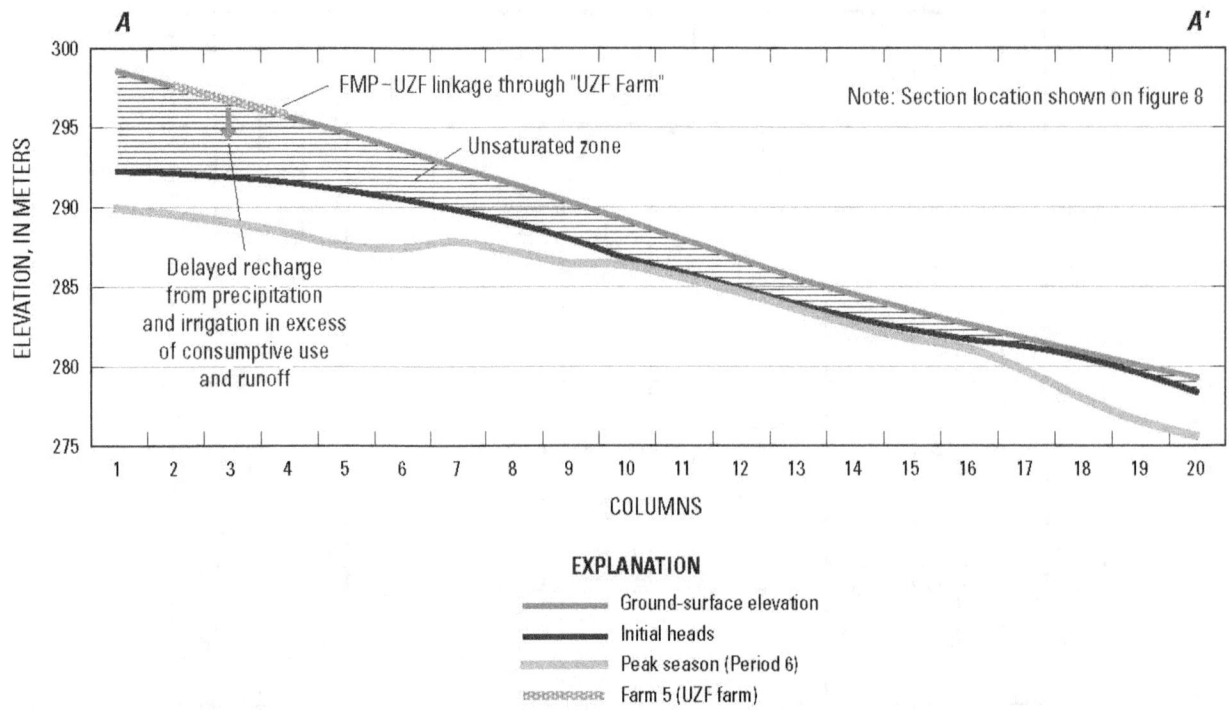

Figure 13.   Relation between the land surface and the water table with an unsaturated zone for the FMP2 example.

Summary and Analysis of Output

The hydrologic budget of the landscape can be analyzed for individual or groups of water-accounting regions (virtual farms). For example, a detailed farm budget includes all physical flows into and out of a farm. Inflow components are precipitation (Q-p-in), evaporation (Q-egw-in) and transpiration (Q-tgw-in) from groundwater uptake , as well as the surface-water (Q-nrd-in, Q-srd-in, Q-rd-in) and groundwater supply (Q-wells-in) components and any potential external deliveries (Q-ext-in) required to balance supply with the simulated demand (example: farm 1, fig. 14

top). Outflow components are evaporation from precipitation, irrigation, and groundwater uptake (Q-ep-out, Q-ei-out, Q-egw-out), transpiration from precipitation, irrigation, and groundwater uptake (Q-tp-out, Q-ti-out, Q-tgw-out), runoff (Q-run-out), and percolation below the root zone (Q-dp-out) (example: farm 1, fig. 14 bottom). Total rates of inflow (Q-tot-in) and outflow (Q-tot-out) also are included with the detailed farm budget components as are cumulative volumes (not shown in figure 14). Alternatively these budget components can be viewed for selected periods of time or individual stress periods or time steps, as shown for farm 1 for the second time step of stress period 6 (fig. 15).

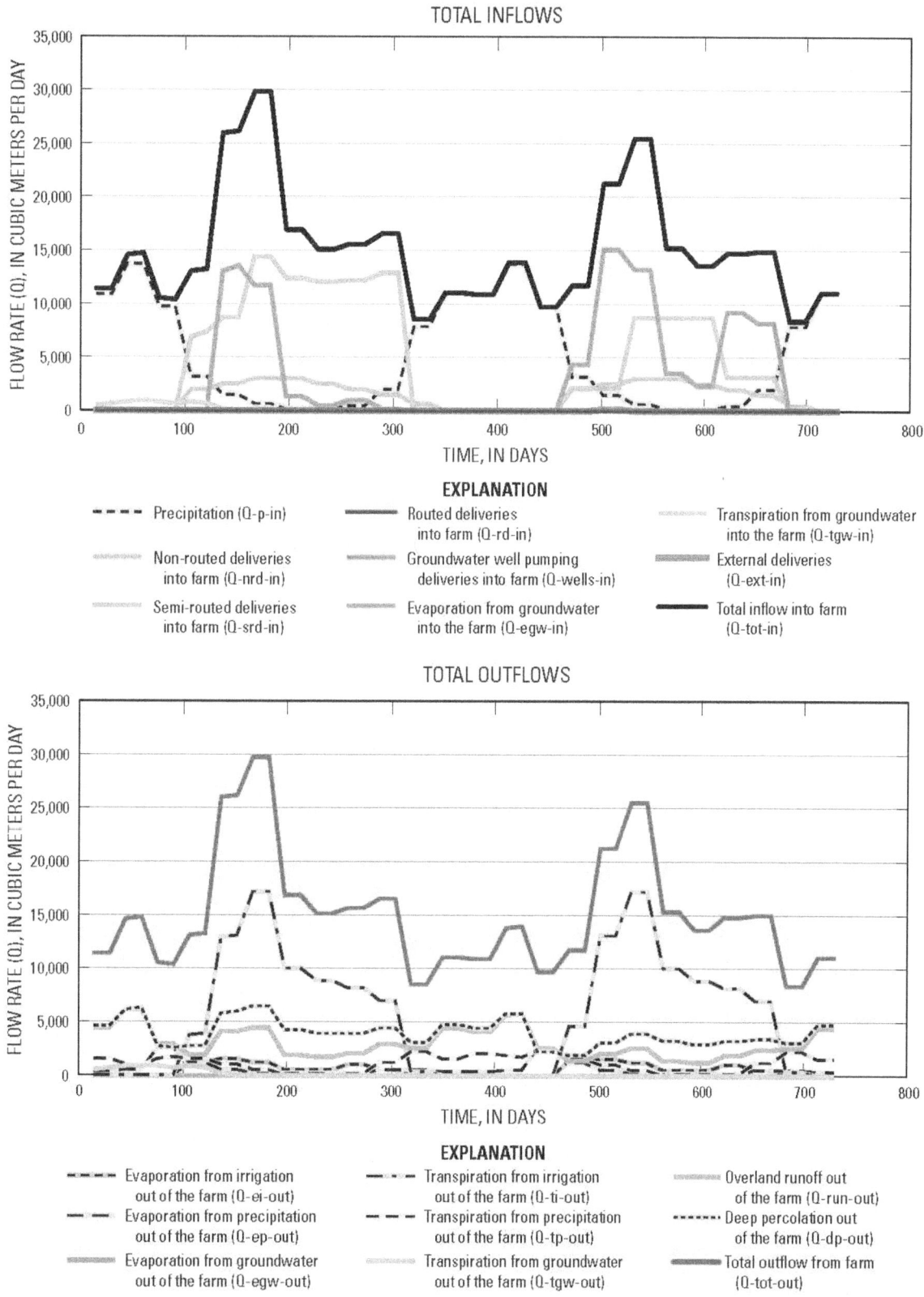

Figure 14. Farm Budget time series for Farm 1. A. Total inflows. B. Total outflows.

Farm budget for:			
Farm:	1	Stress period: 6	Time step: 2
Volumetric Flowrates [L³/T, here meters cubed/day]			
INFLOWS:			
Precipitation		Q-p-in	682.50
Non-routed deliveries into farm		Q-nrd-in	2,958.58
(Semi-)Routed deliveries from remote canal		Q-srd-in	14,346.95
Fully routed deliveries from adjacent canal		Q-rd-in	0.00
Farm well pumping		Q-wells-in	11,793.24
Evaporation from groundwater		Q-egw-in	0.00
Transpiration from groundwater		Q-tgw-in	0.00
External deliveries		Q-ext-in	0.00
TOTAL INFLOWS TO FARM:		TOTAL IN	29,781.26
OUTFLOWS:			
Evaporation from irrigation		Q-ei-out	1,204.40
Evaporation from precipitation		Q-ep-out	228.74
Evaporation from groundwater		Q-egw-out	0.00
Transpiration from irrigation		Q-ti-out	17,111.48
Transpiration from precipitation		Q-tp-out	453.77
Transpiration from groundwater		Q-tgw-out	0.00
Surface-water runoff		Q-run-out	4,366.67
Deep percolation		Q-dp-out	6,416.21
Non-routed deliveries out of farm		Q-nrd-out	0.00
Excess non-routed deliveries returned to remote canal		Q-srd-out	0.00
Excess non-routed deliveries returned to adjacent canal		Q-rd-out	0.00
Excess non-routed deliveries injected into farm-wells		Q-wells-out	0.00
TOTAL OUTFLOWS OUT OF FARM:		TOTAL OUT	29,781.26
IN - OUT:			0.00
DISCREPANCY, IN PERCENT:			0.00

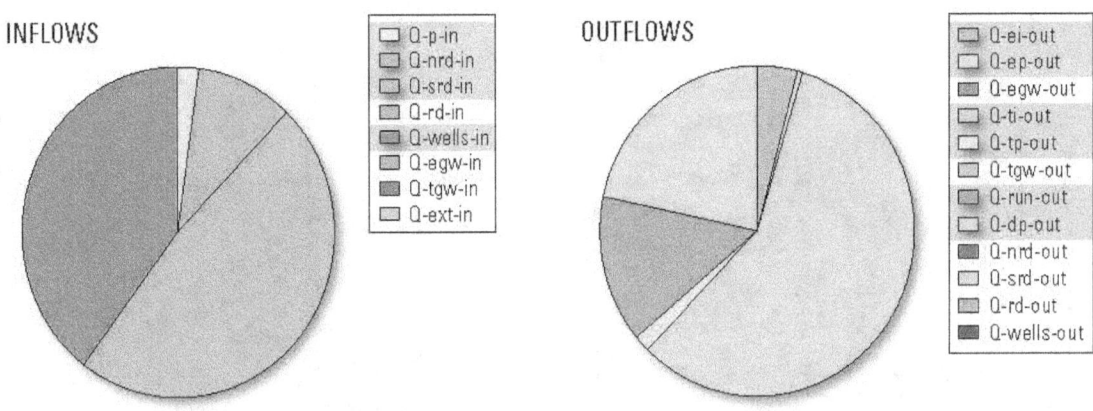

Highlighted components are nonzero

Figure 15.   Farm Budget for Farm 1 for a specific stress period and time step.

The Farm Demand and Supply budget components can be viewed for scenarios of assumed sufficiency (zero scenario, IDEFFL = 0) where potential model-internal deficiencies are assumed to be balanced by model-external "External Deliveries" (example: farm 6, fig. 16A). Farm Demand and Supply budgets also can be viewed for deficiency scenarios where the initial supply and demand rates reflect the initial assessment of resources and irrigation demands at the beginning of each time step and where the final components show how the irrigation demand was reduced to fit the available supply. For the deficit irrigation scenario (IDEFFL = –1), this reduction is achieved in the model first by improving the on-farm efficiency, if necessary to 100 percent, and second by reducing the crop consumptive use such that the transpiration and evaporation from irrigation cannot exceed the supplied irrigation (example: farm 6, fig. 16B). By running this scenario, one can simulate the reduced actual crop evapotranspiration, returnflows, and deep percolation that follow droughts. This, in turn allows the derivation of reduced crop yields. Another deficiency scenario in FMP, called "Acreage Optimization" (IDEFFL ≥1), optimizes a farm's profit against irrigated acreage once a farm experiences insufficient supply. The Acreage Optimization calculates yield, benefit, and profit associated with a reduced acreage code-internally. The Acreage-Optimization scenario is not part of the present example model but was previously demonstrated in the FMP1 user guide (Schmid and others, 2006).

This example demonstrates how the equal-appropriation allotment limit (ALLOT) on semi-routed surface-water deliveries creates a demand for groundwater pumpage followed by reaching a pumping capacity limit which creates an irrigation deficiency in the second year (fig. 16A,B). Alternatively these budget components can be viewed for a specific stress period (17) for farm 6 to analyze the particulars of the deficiency and related operational drought resulting from an insufficient surface-water and groundwater supplies during the summer of the second year of the simulation (fig. 17). In this example the reduction of demand by about 27 percent is achieved (fig. 17). For farm 6, the pumping capacity related to two wells is limited in one well by its drawdown reaching a head constraint (fig. 18A) set in the linked MNW data input file before the maximum pumping

rate set in FMP2 was met (Appendix A, file ex1_MN.mnw) and in another well by meeting the maximum pumping-rate constraint (fig. 18B) set in the FMP2 well list (Appendix A, file WELLS.IN) .

Supplemental groundwater is pumped from both FMP2-internal farm wells and MNW wells externally linked to FMP2 farm wells. The link between FMP2 and MNW enables the simulation of (a) the apportioning of FMP-calculated desired flow rates, (b) actual flow rates, and (c) hydraulically determined maximum capacities of multi-layer wells. The connection to MNW can facilitate additional constraints (fig. 1), such as head or drawdown constraints that can keep the pumping level in the cell above the bottom of the well, bottom of the aquifer, or above some other important threshold related to other processes such as land subsidence or seawater intrusion.

Delayed recharge is facilitated in the example model by linking the percolation beyond the root zone to the applied infiltration of the UZF Package, which simulates runoff in excess of the actual infiltration, recharge, and for conditions when water levels rise above the surface, discharge of groundwater to surface water and rejected infiltration.

In the example model, both delayed recharge and infiltration in excess of the saturated hydraulic conductivity occur beneath the "UZF" farm at cell (3, 3) in farm 5 (figs. 8 and 12). The delayed recharge occurs about 160 days after the onset of percolation of the farm process in the first year (fig.19A). The time lag between the maxima of percolation (applied infiltration in UZF1) and of recharge is about 240 days. Runoff from infiltration in excess of the saturated hydraulic conductivity occurs during the peak irrigation seasons of both years. This runoff is an additional component to the FMP2-generated runoff created by consumptive-use-inefficient losses and is added to the total runoff tabulated in the detailed farm budget of FMP2 (if IFBPFL = 2 is set). All or part of the actual infiltration is taken into vadose zone storage during the springs and summers of both years because of heavy winter–spring precipitation and heavy peak season irrigation during the summer. During the winters of both years, water is released again from the vadose zone storage and contributes vertically downward to the delayed recharge (fig. 19A).

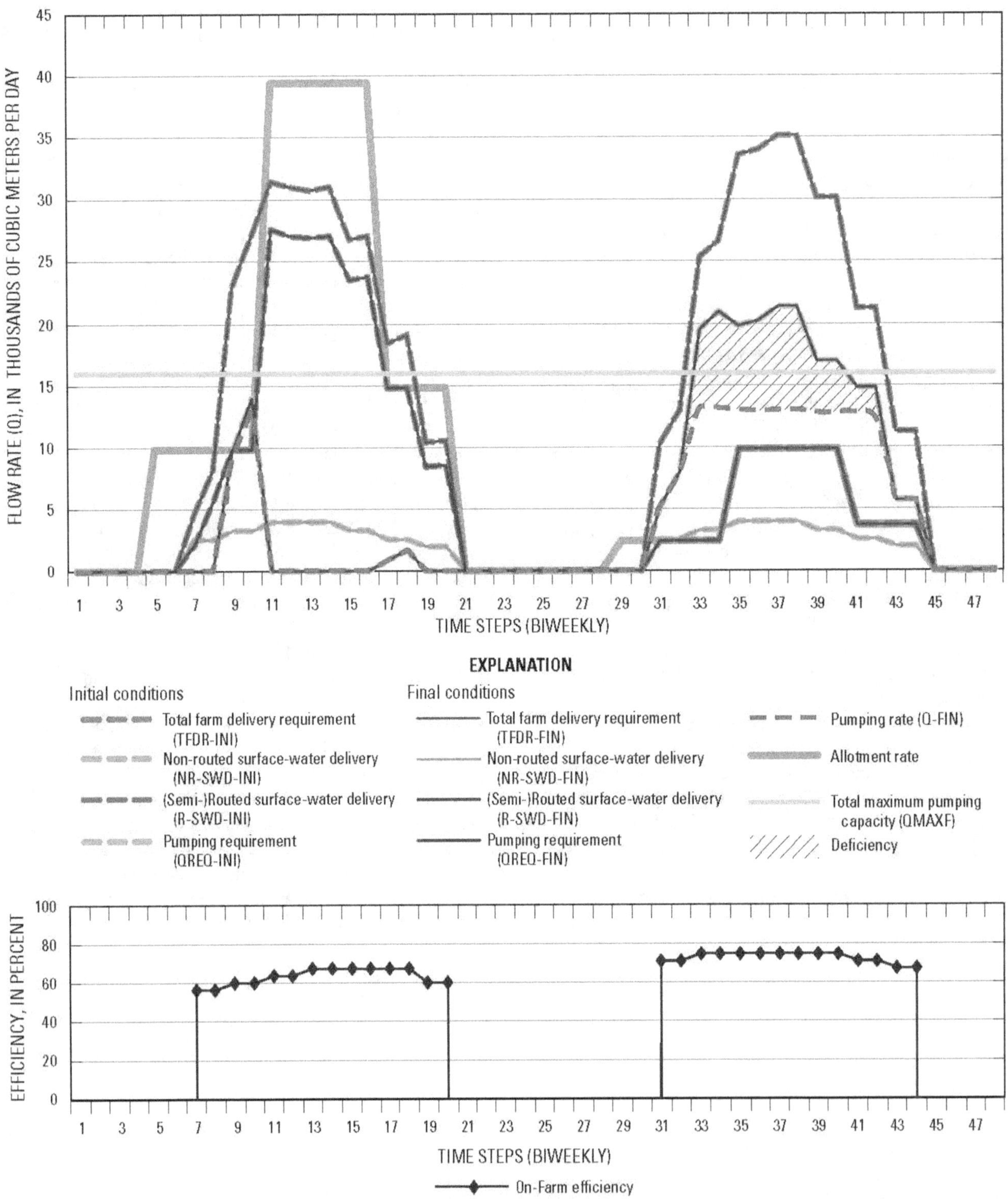

Figure 16. Simulated Farm Demand and Supply components and simulated on-farm efficiency for Farm 6 (urban area) (*A*) assuming a zero scenario for deficiency situations, and (*B*) deficit irrigation scenario for Farm 6 in response to a deficiency situation with initial and final Farm demand and supply components.

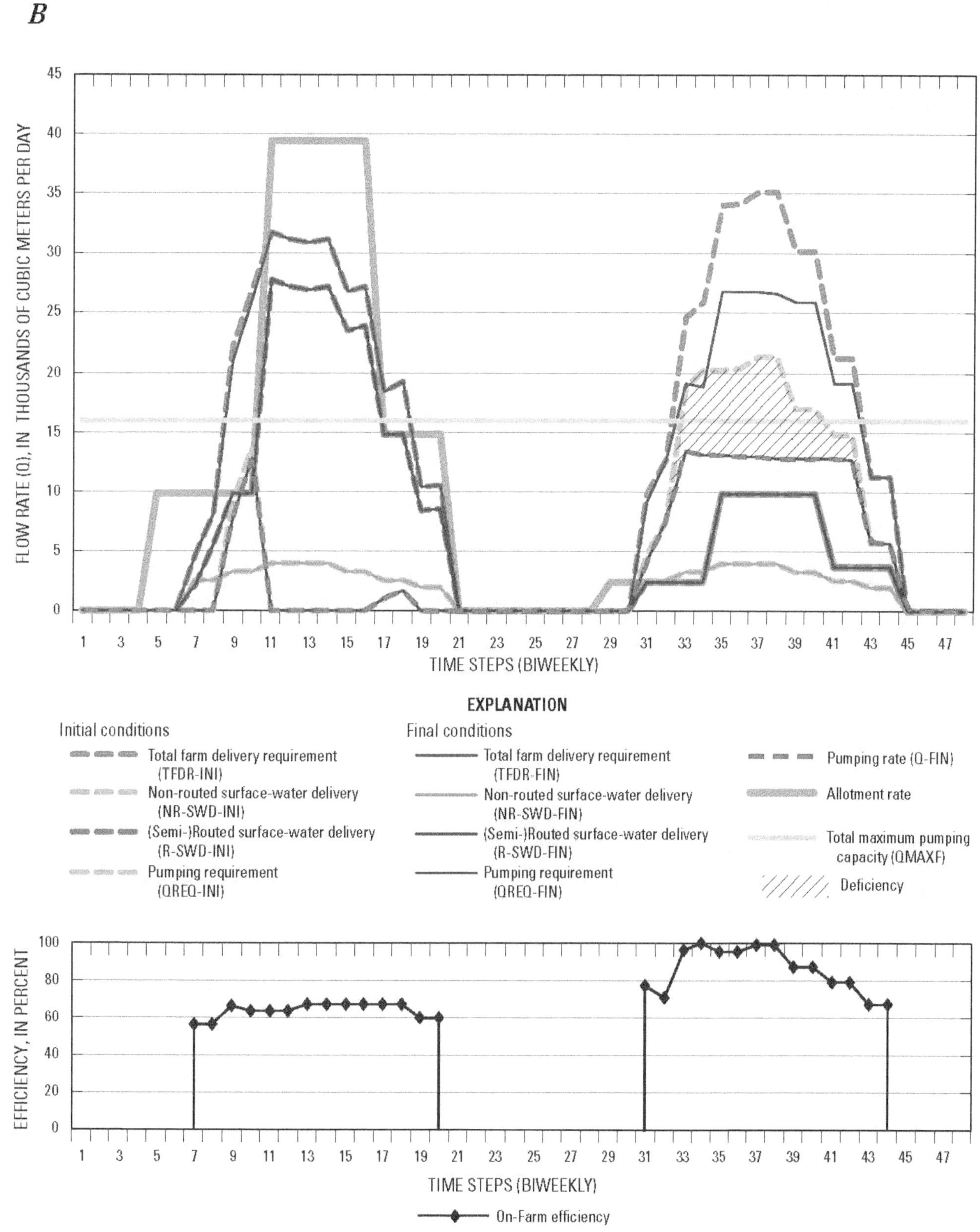

B
FLOW RATE (Q), IN THOUSANDS OF CUBIC METERS PER DAY
0
5
10
15
20
25
30
35
40
45
1 3 5 7 9 11 13 15 17 19 21 23 25 27 29 31 33 35 37 39 41 43 45 47
TIME STEPS (BIWEEKLY)
EXPLANATION
Initial conditions
Total farm delivery requirement (TFDR-INI)
Non-routed surface-water delivery (NR-SWD-INI)
(Semi-)Routed surface-water delivery (R-SWD-INI)
Pumping requirement (QREQ-INI)
Final conditions
Total farm delivery requirement (TFDR-FIN)
Non-routed surface-water delivery (NR-SWD-FIN)
(Semi-)Routed surface-water delivery (R-SWD-FIN)
Pumping requirement (QREQ-FIN)
Pumping rate (Q-FIN)
Allotment rate
Total maximum pumping capacity (QMAXF)
Deficiency
EFFICIENCY, IN PERCENT
0
20
40
60
80
100
1 3 5 7 9 11 13 15 17 19 21 23 25 27 29 31 33 35 37 39 41 43 45 47
TIME STEPS (BIWEEKLY)
On-Farm efficiency

Figure 16.—Continued.

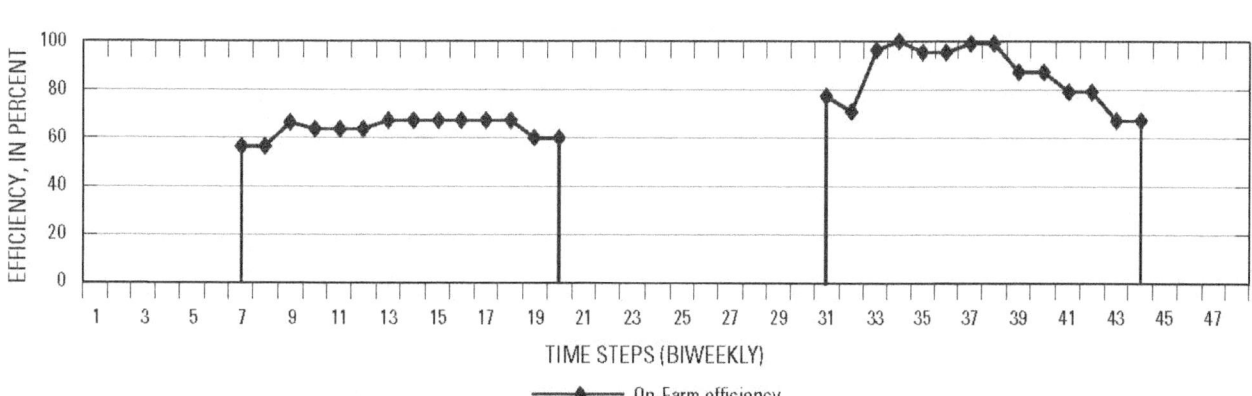

Farm Demand and Supply Budget for:				
Farm: 6	Stress period: 17	Time step: 2		

INITIAL RATES (before deficiency response scenario):		Demand:	Supply:	Demand − Supply
Total farm delivery requirement	TFDR-INI	25,931.81		
Non-routed deliveries to farm	NR-SWD-INI		3,287.31	
(Semi-)Routed deliveries	R-SWD-INI		2,465.48	
Available farm well pumping	min{QREQ-INI; QMAXF}		13,097.63	
TFDR-INI − NRD-INI − {S}RD-INI − min{QREQ-INI;QMAXF}				7,081.38

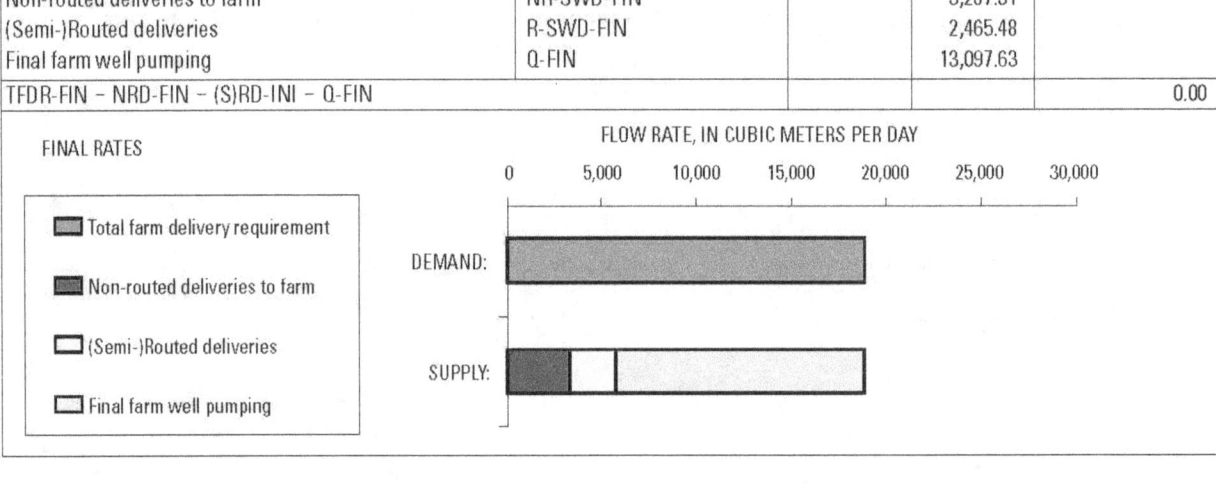

FINAL RATES (after deficiency response scenario):		Demand:	Supply:	Demand − Supply
Deficit irrigation				
Total farm delivery requirement	TFRD-FIN	18,850.43		
Non-routed deliveries to farm	NR-SWD-FIN		3,287.31	
(Semi-)Routed deliveries	R-SWD-FIN		2,465.48	
Final farm well pumping	Q-FIN		13,097.63	
TFDR-FIN − NRD-FIN − {S}RD-INI − Q-FIN				0.00

Figure 17. Effect of deficit irrigation scenario for Farm 6 in response to a deficiency situation with initial and final Farm demand and supply components.

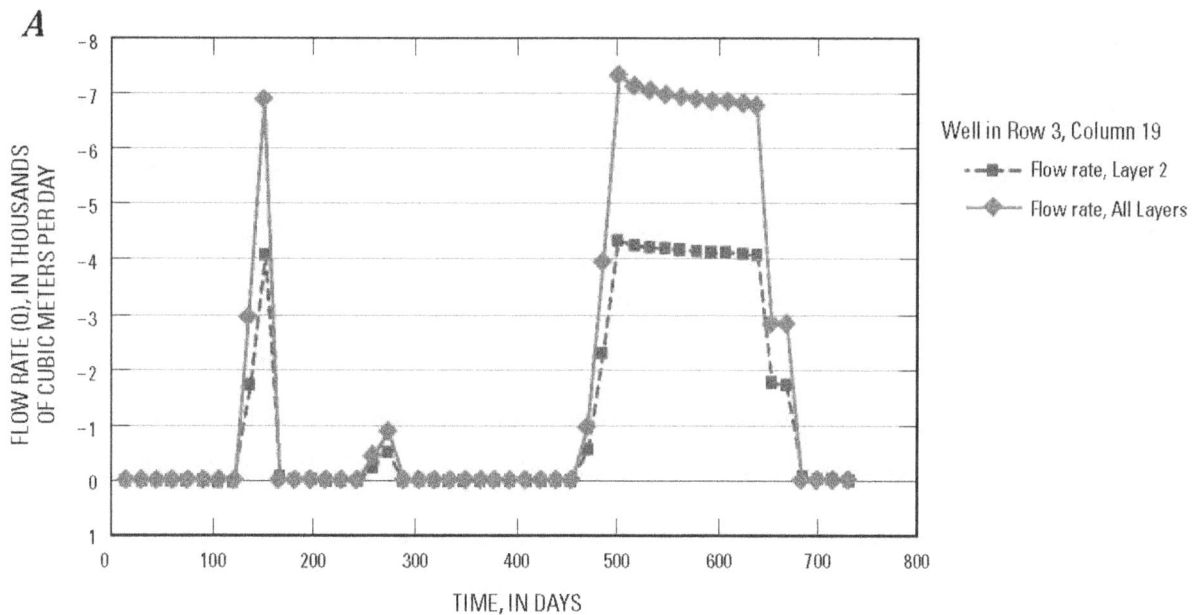

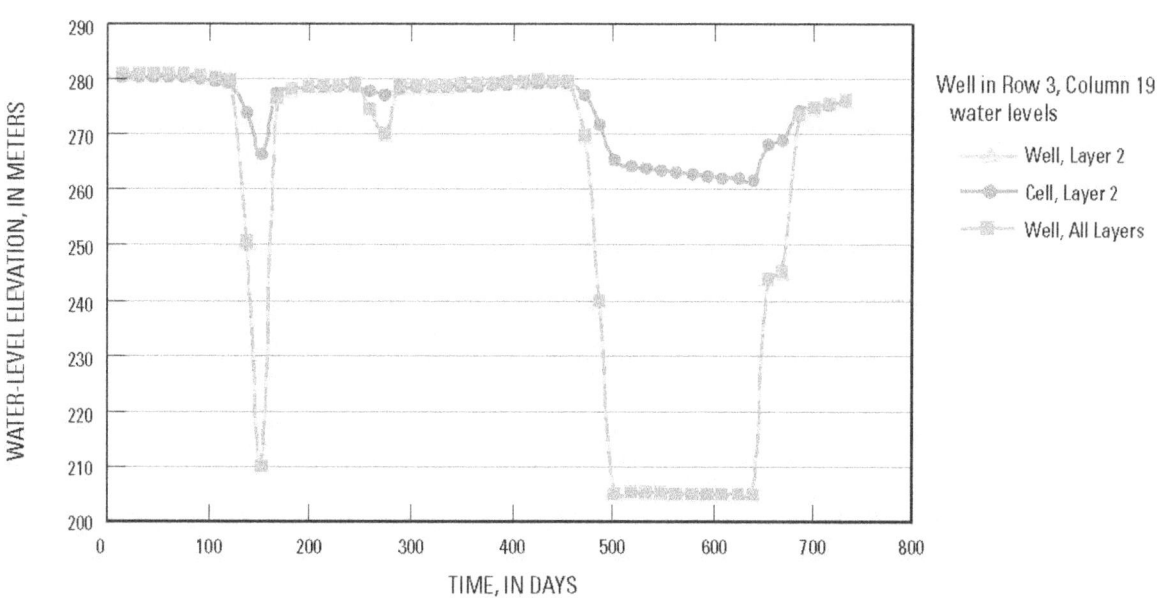

Figure 18. Response of a Multi-node Farm well to demand and supply and to head-constraint set in MNW for individual layer pumpage and net pumpage of the wells (*A*) in row 3, column 19, and (*B*) in row 6, column 17 of Farm 6 in the FMP2 example.

B

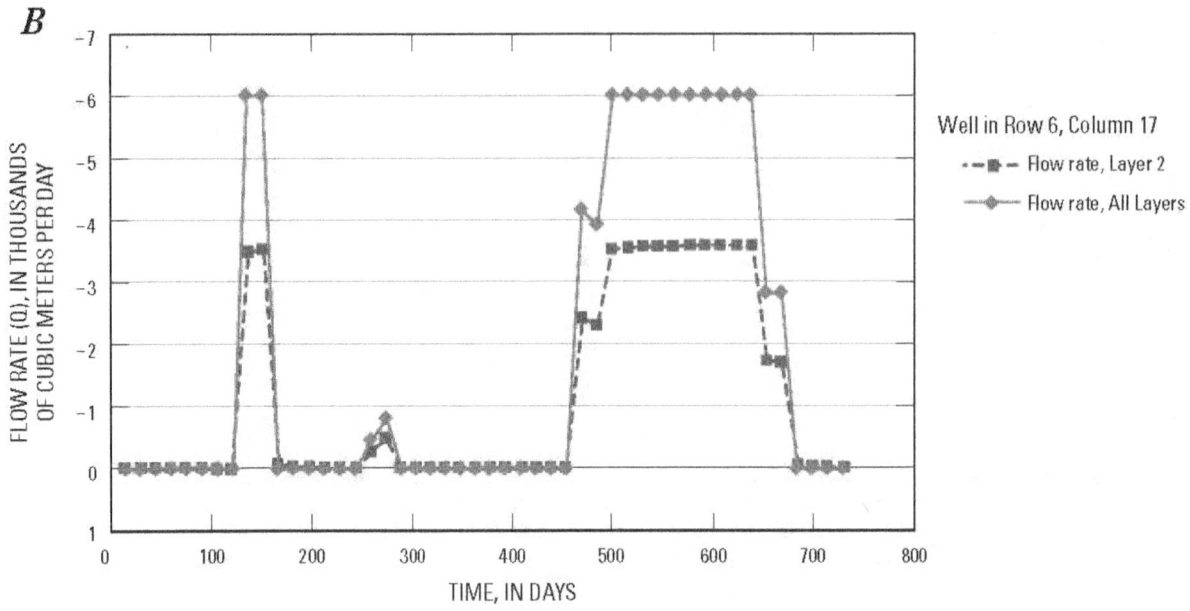

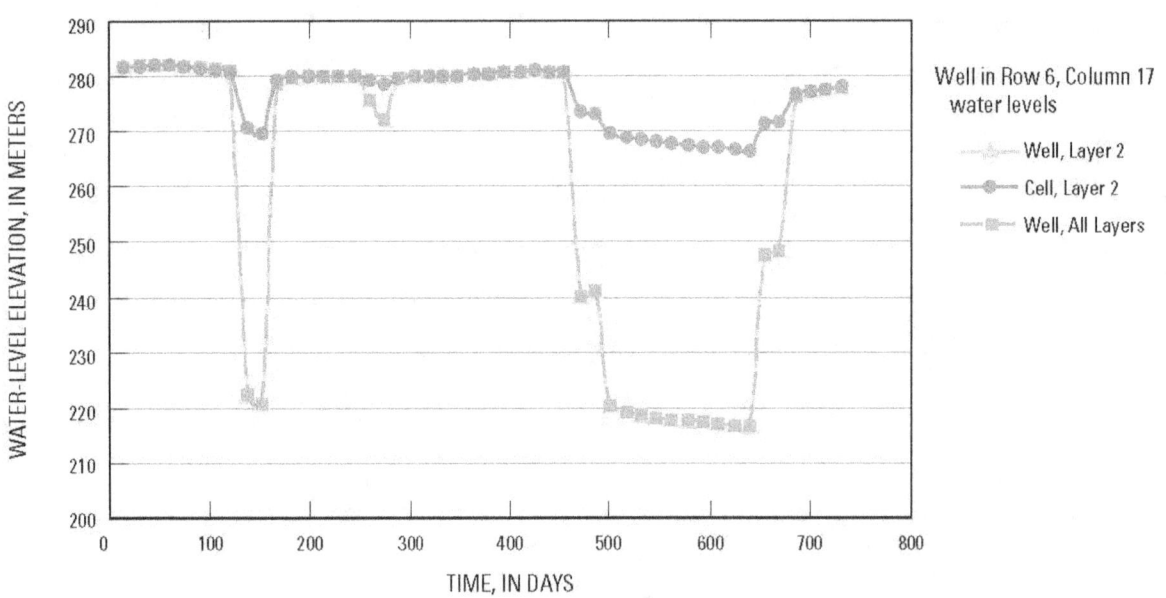

Figure 18.—Continued.

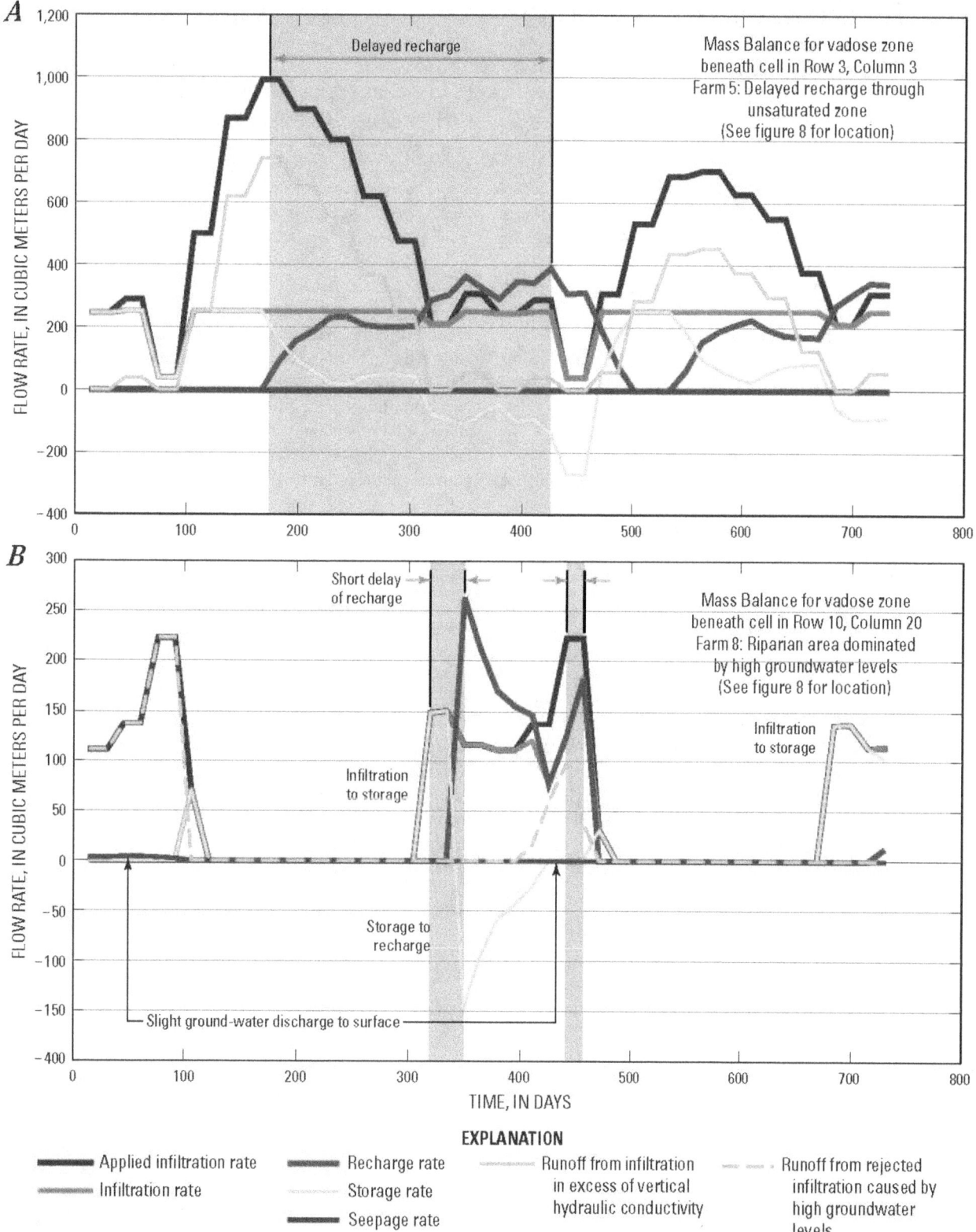

Figure 19. Delayed recharge and runoff from infiltration in excess of the saturated vertical hydraulic conductivity in (*A*) cell (3,3) at Farm 5 that overlies an unsaturated zone in the example model, and (*B*) the groundwater discharge to surface and runoff from rejected infiltration caused by high groundwater levels in cell (10,20) at Farm 8 in the example model.

In contrast to farm 5, which is characterized by a deep vadose zone, the riparian area (farm 8) is dominated by high groundwater levels that initially rise slightly above the ground surface. For the riparian area, the UZF1-FMP2 linkage can simulate groundwater discharged to the surface during times of heavy winter–spring precipitation typical for the climate of Davis, California, and under these conditions of high groundwater levels, runoff from rejected infiltration (fig 19B). In the absence of summer precipitation and irrigation, willow trees take up water under saturated and unsaturated conditions, which results in a water-level decline over each year. After the onset of new winter precipitation, actual infiltration beyond the root zone is taken into unsaturated storage and, because of a shallow unsaturated zone, shortly thereafter again released from storage to recharge (fig. 19B). The delay between percolation and recharge is only about 40 days for the riparian area, compared to the 240-day delay for farm 5, where a deep vadose zone explains the longer delay (fig. 19B).

Simulating delayed recharge and additional runoff components (from infiltration in excess of the saturated hydraulic conductivity, from groundwater discharge to surface water, and from rejected infiltration under conditions of high groundwater levels) provides additional coupling for complex conjunctive use scenarios through the linkage of FMP2 with UZF1 in MF-FMP.

The mutual effect that the linked flow terms have on one another as conjunctive use was assessed using the physical budget and demand and supply budget of virtual farms representing surface landscape water-accounting units (figs. 14–17) and through subsurface groundwater accounting units using MODFLOW's ZONEBUDGET (Harbaugh, 1990). In the example, the groundwater accounting units (zones) are delineated equally to the surface accounting units (farms) that overlie them (Appendix A, files FMPzone.in and FMPzone.zon). For each zone, or in this case farm, boundary flows and flows from surrounding other farms into and out of a farm are calculated by ZONEBUDGET from the cell-by-cell flows written to a 'compact budget' binary file. The use of ZONEBUDGET in conjunction with FMP2 requires the specification of "COMPACT BUDGET" in the Output Control File. Pie charts of groundwater budgets for particular zones and time frames (farm 5, fig. 20) or time series of groundwater budgets for each or all zones (fig. 21) can be derived from this ZONEBUDGET data output.

The example model also used HYDMOD (Hanson and Leake, 1998) (Appendix A, file hyd_all.hyd) to generate hydrographs of groundwater levels in select observation wells and streamflow hydrographs of inflows and outflows at select reaches of the streamflow routing network. In the example, the reaches coincide with locations where diversion to, and returnflows from, farms occur. However, using HYDMOD other random reaches along the streamflow network could be chosen to investigate changes in streamflow. The difference between inflows and outflows at the diversion reaches allows an easy and synoptic view of the diversion to the respective farms (fig. 22A). However, the difference between inflows and outflows at the returnflow reaches shows the net changes in streamflow not only caused by the runoff returnflow from a particular farm, but also related to prorated runoff returnflow from the native vegetation (farm 7) or from the riparian area (farm 8) (fig. 22B). In addition, streambed leakage might influence changes in streamflow between inflows and outflows at a particular returnflow reach to a minor degree. Runoff to any returnflow reach is accounted for cumulatively for several sources such as farms, native and riparian vegetation, and urban areas. HYDMOD in conjunction with MF2005-FMP2 provides a tool to evaluate cumulative surface-water returnflows at any point along the stream network. One real world application of this option is to allow water managers of different states participating in a stream compact to evaluate the effect of cumulative returnflows of a stream prior to reaching a state boundary.

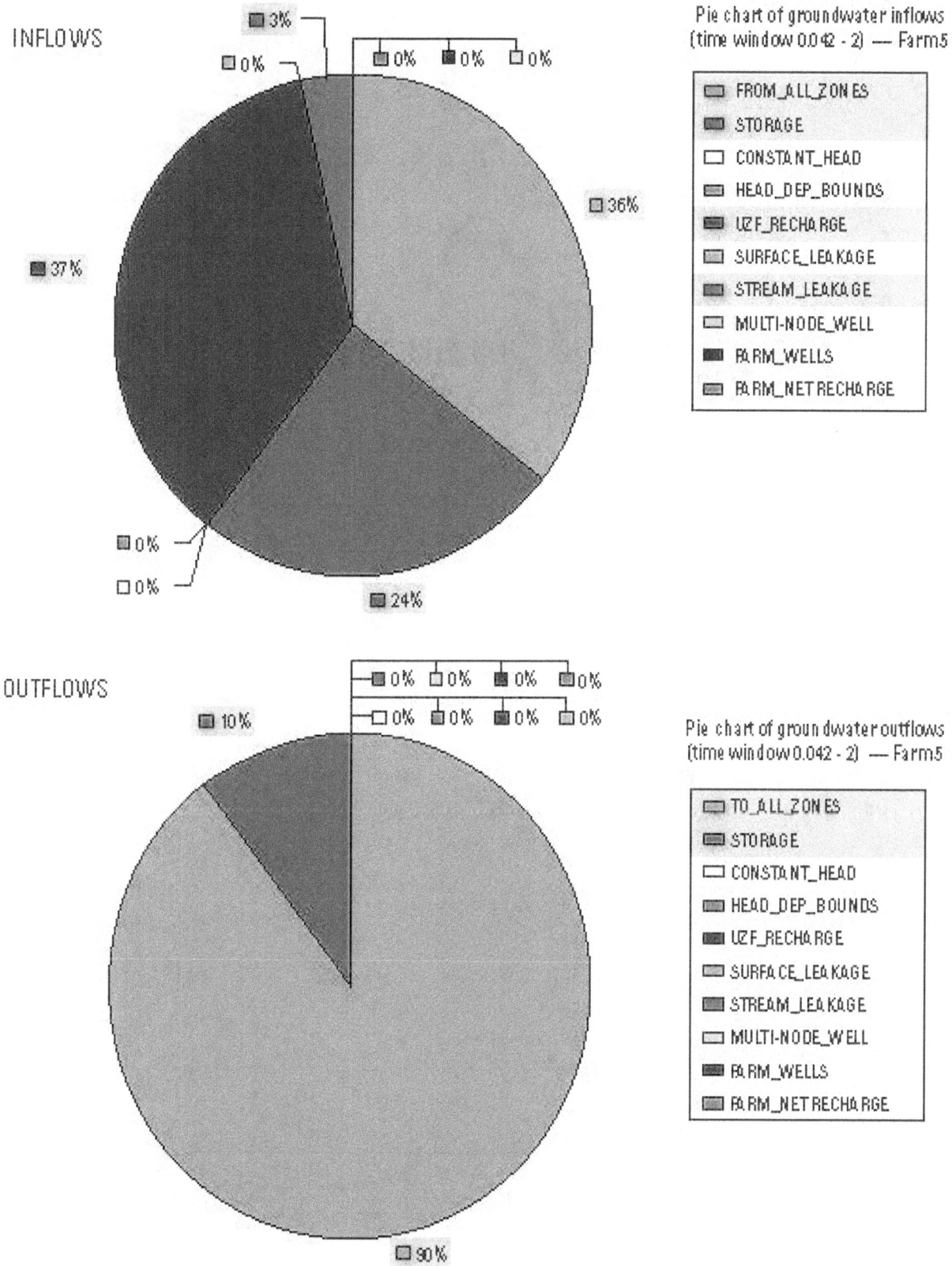

INFLOWS

3%   0%   0%   0%

0%

36%

37%

0%
0%

24%

Pie chart of groundwater inflows
(time window 0.042 - 2) — Farm5

- FROM_ALL_ZONES
- STORAGE
- CONSTANT_HEAD
- HEAD_DEP_BOUNDS
- UZF_RECHARGE
- SURFACE_LEAKAGE
- STREAM_LEAKAGE
- MULTI-NODE_WELL
- FARM_WELLS
- FARM_NET RECHARGE

OUTFLOWS

0%   0%   0%   0%
0%   0%   0%   0%

10%

90%

Pie chart of groundwater outflows
(time window 0.042 - 2) — Farm5

- TO_ALL_ZONES
- STORAGE
- CONSTANT_HEAD
- HEAD_DEP_BOUNDS
- UZF_RECHARGE
- SURFACE_LEAKAGE
- STREAM_LEAKAGE
- MULTI-NODE_WELL
- FARM_WELLS
- FARM_NET RECHARGE

Figure 20.   Groundwater budget components derived by ZONEBUDGET for Farm 5 for specific periods.

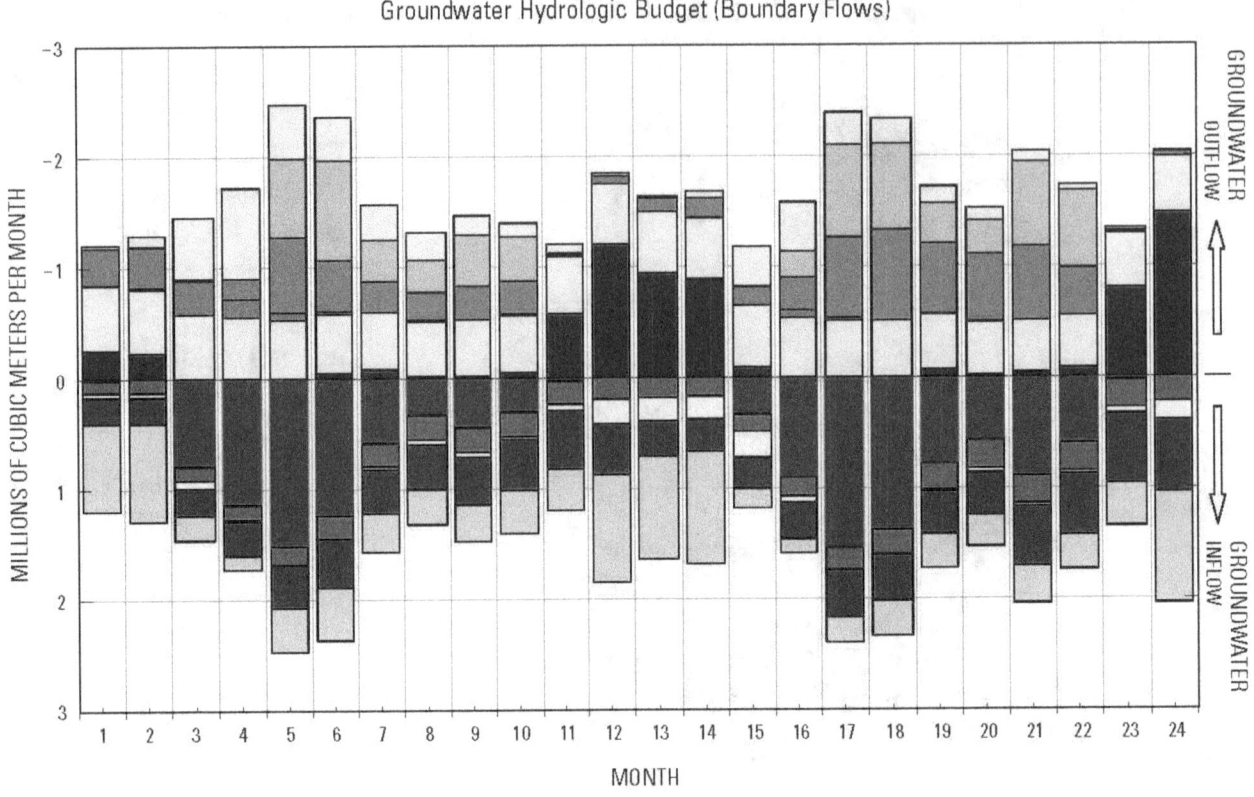

Figure 21. Cumulative groundwater budget components for all Farms throughout the simulation period.

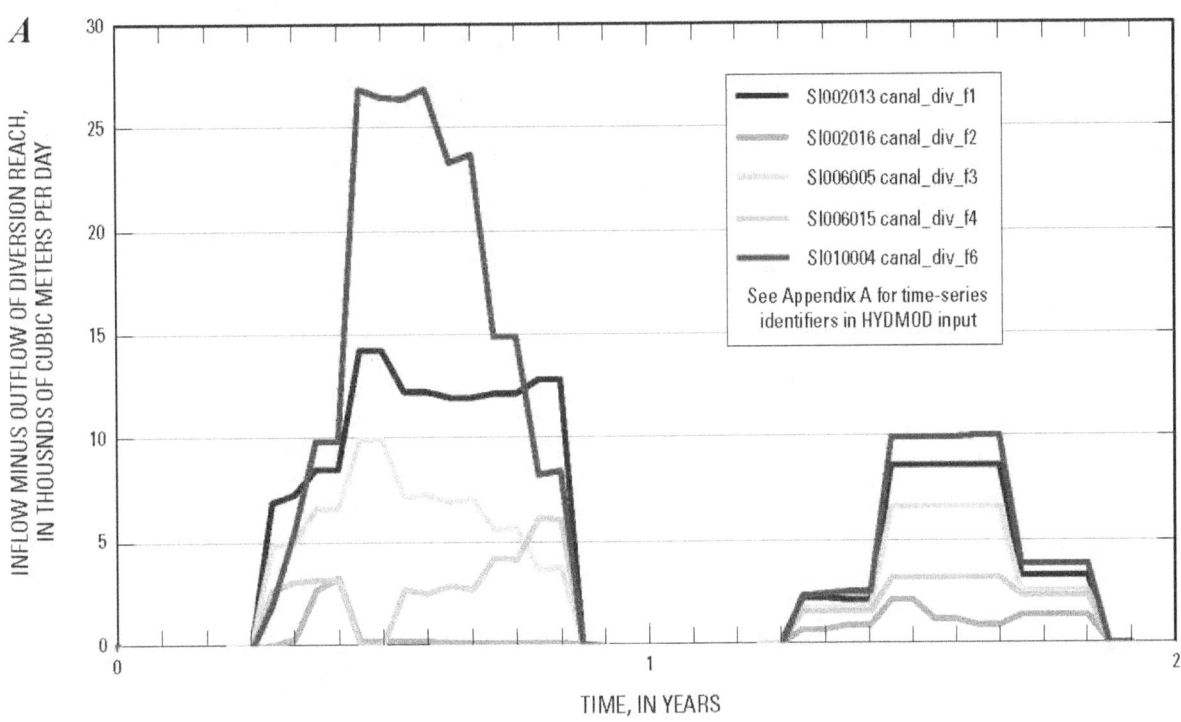

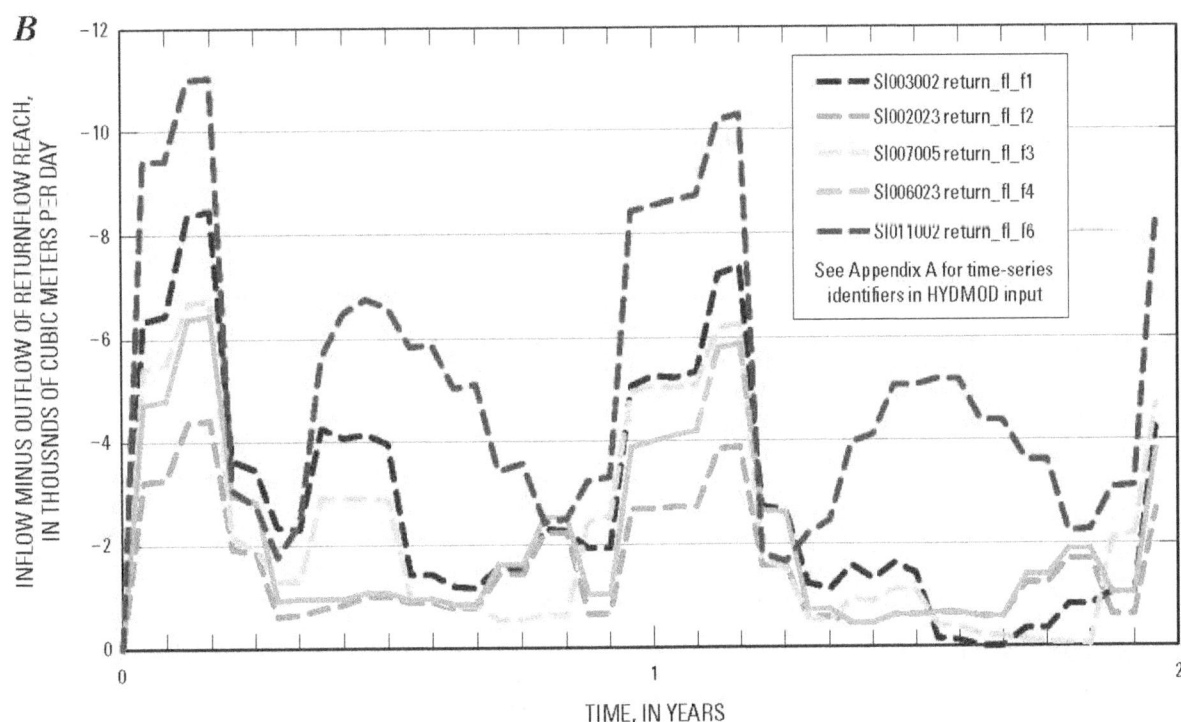

Figure 22. Changes in streamflow at points of diversion to farms (*A*) and points of returnflow from farms (*B*) derived from HYDMOD output data.

References

Allen, R.G., Pereira, L.S., Raes, D., and Smith, M., 1998, Crop evapotranspiration — guidelines for computing crop water requirements: Food and Agriculture Organization of the United Nations, Irrigation and Drainage Paper 56, 300 p.

Allen, R.G., Clemmens, A.J., Burt, C.M., Solomon, K., and O'Halloran, T., 2005, Prediction Accuracy for Projectwide Evapotranspiration Using Crop Coefficients and Reference Evapotranspiration, Journal of Irrigation and Drainage Engineering, ASCE, v.. 131, no. 1, p. 24-36.

Belitz, K., Phillips, S.P., and Gronberg, J.M., 1993, Numerical simulation of ground-water flow in the central part of the western San Joaquin Valley, California: U.S. Geological Survey Water-Supply Paper 2396, 69 p.

Brooks, R.H., and Corey, A.T., 1966, Properties of porous media affecting fluid flow: American Society of Civil Engineers, Journal of Irrigation and Drainage, v. 101, p. 85–92.

Brush, C.F., Belitz, Kenneth, and Phillips, S.P., 2004, Estimation of a water budget for 1972–2000 for the Grasslands Area, Central Part of the Western San Joaquin Valley, California: U.S. Geological Survey Scientific Investigations Report 2004–5180, 51 p.

Food and Agriculture Organization, 2007, Chapter 8 - ET$_c$ under soil water stress conditions: accessed September 9, 2008, at http://www.fao.org/docrep/x0490e/x0490e0e.htm

Faunt, C.C., Hanson, R.T., and Belitz, K., 2008a, Development of a Model to Assess Ground-water Availability in California's Central Valley: Water Resources IMPACT, American Water Resources Association, v. 10, no. 1, p. 27–30.

Faunt, C.C., Hanson, R.T., Schmid, W., Belitz, K., 2008b, Application of MODFLOW's Farm Process to California's Central Valley: Modflow and More – Ground Water and Public Policy, Golden, Colorado, May 18-21, 2008, p. 496–500.

Faunt, C.C., Hanson, R.T., Belitz, K., 2009a, Chapter A. Introduction, Overview of Hydrogeology, and Textural Model of California's Central Valley, in Faunt, C.C. ed., Groundwater Availability of the Central Valley Aquifer, California: U.S. Geol. Survey Professional Paper 1776, p. 1-58.

Faunt, C.C., Belitz, K., and Hanson R.T., 2009b, Chapter B. Groundwater availability in California's Central Valley, in Faunt, C.C. ed., Groundwater Availability of the Central Valley Aquifer, California: U.S. Geological Survey Professional Paper 1776, p. 59-120.

Faunt, C.C., Hanson, R.T., Belitz, K., Schmid, W., Predmore, S.P., Rewis, D.L., McPherson, K.R., 2009c, Chapter C. Numerical Model of the Hydrologic Landscape and Groundwater Flow in California's Central Valley, in Faunt, C.C. ed., Groundwater Availability of the Central Valley Aquifer, California: U.S. Geological Survey Professional Paper 1776, p. 121-212.

Halford, K.J. and Hanson, R.T., 2002, User guide for the drawdown- limited, multi-node well (MNW) package for the U.S. Geological Survey's modular three-dimensional finite-difference ground-water flow model, versions MODFLOW-96 and MODFLOW-2000: U.S. Geological Survey Open-File Report 02-293, 33 p.

Hanson, R.T., and Leake, S.A., 1998, Documentation for HYDMOD, A program for time-series data from the U.S. Geological Survey's modular three-dimensional finite-difference ground-water flow model: U.S. Geological Survey Open-File Report 98-564, 57 p., accessed April 17, 2009, at http://pubs.er.usgs.gov/usgspubs/ofr/ofr98564

Hanson, R.T., Schmid, W. and Leake, S.A., 2008a, Assessment of Conjunctive Use Water-Supply Components Using Linked Packages and Processes in MODFLOW: Modflow and More – Ground Water and Public Policy, Golden, Colorado, May 18–21, 2008, p. 5.

Hanson, R.T., Schmid, W., Lear, J., and Faunt, C.C., 2008b, Simulation of an Aquifer-Storage-and-Recovery (ASR) System using the Farm Process in MODFLOW for the Pajaro Valley, Monterey Bay, California: Modflow and More – Ground Water and Public Policy, Golden, Colorado, May 18-21, 2008, p. 501–505.

Harbaugh, A.W., 1990, A computer program for calculating subregional water budgets using results from the U.S. Geological Survey Modular Three-dimensional Finite-difference Ground-water Flow Model: U.S. Geological Survey Open-File Report 90-392, 24 p.

Harbaugh, A.W., 2005, MODFLOW-2005, the U.S. Geological Survey modular ground-water model—the Ground-Water Flow Process: U.S. Geological Survey Techniques and Methods 6-A16, variously paged.

Harbaugh, A.W., Banta, E.R., Hill, M.C., and McDonald, M.G., 2000, MODFLOW-2000: U.S. Geological Survey Modular Ground-Water Model—User Guide to Modularization Concepts and the Ground-Water Flow Process: U.S. Geological Survey Open-File Report 00-92, 121 p.

Hoffmann, J., Leake, S.A., Galloway, D.L., and Wilson, A.M., 2003, MODFLOW-2000 ground-water model—user guide to the subsidence and aquifer-system compaction (SUB) package: U.S. Geological Survey Open-File Report 03–233, 46 p.

McAda, D.P., and Barroll, P., 2002, Simulation of ground-water flow in the middle Rio Grande Basin between Cochiti and San Acacia, New Mexico: U.S. Geological Survey Water-Resources Investigations Report 02-4200, 88 p.

Merritt, M.L., and Konikow, L.F., 2000, Documentation of a computer program to simulate lake-aquifer interaction using the MODFLOW ground-water flow model and the MOC3D solute-transport model: Water-Resources Investigations Report 00-4167, 146 p.

Niswonger, R.G., and Prudic, D.E., 2005, Documentation of the Streamflow-Routing (SFR2) Package to include unsaturated flow beneath streams — A modification to SFR1: U.S. Geological Survey Techniques and Methods 6-A13.

Niswonger, R.G., Prudic, D.E., and Regan, R.S., 2006, Documentation of the Unsaturated-Zone Flow (UZF1) Package for modeling unsaturated flow between the land surface and the water table with MODFLOW-2005: U.S. Geological Survey Techniques and Methods 6-A19, 62 p.

Prudic, D.E., Konikow, L.F., and Banta, E.A., 2004, A new Streamflow-Routing (SFR1) Package to simulate stream-aquifer interaction with MODFLOW-2000: U.S. Geological Survey Open-File Report 04-1042, 95 p.

Schmid, W., 2004, A Farm Package for MODFLOW-2000: Simulation of Irrigation Demand and Conjunctively Managed Surface-Water and Ground-Water Supply; PhD Dissertation: Department of Hydrology and Water Resources, The University of Arizona, 278 p.

Schmid, W., Hanson, R.T., Maddock III, T.M., and Leake, S.A., 2006, User's guide for the Farm process (FMP) for the U.S. Geological Survey's modular three-dimensional finite-difference ground-water flow model, MODFLOW-2000: U.S. Geological Survey Techniques and Methods 6-A17, 127 p.

Schmid, W., Hanson, R.T., Faunt, C.C. and Phillips, S.P., 2008, Hindcast of water availability in regional aquifer systems using MODFLOW's Farm Process: Proceedings of Hydropredict 2008, Prague, Czech Republic, September 15–19, 2008, pp. 311-314.

Schmid, W., and Hanson, R.T., 2009, Appendix 1, Supplemental Information—Modifications to Modflow-2000 Packages and Processes, *in* Faunt, C.C. ed., Groundwater Availability of the Central Valley Aquifer, California: U.S. Geological Survey Professional Paper 1776, p. 213-225.

Smith, R.E., 1983, Approximate sediment water movement by kinematic characteristics: Soil Science Society of America Journal, v. 47, p. 3-8.

Snyder, R.L., and Eching, S., 2007, Urban landscape evapotranspiration: California Water Plan Update 2005: California Department of Water Resources, accessed September 9, 2008, at http://www.waterplan.water.ca.gov/docs/cwpu2005/vol4/vol4-landscapewateruse-urbanlandscapeevapotranspiration.pdf

Taylor, S.A., and G.M. Ashcroft. 1972, Physical edaphology: the physics of irrigated and non-irrigated soils, San Francisco, Freeman and Co., p. 434-435, 533 p.

Wesseling, J.G., 1991, Meerjarige simulaties van grondwateronttrekking voor verschillende bodemprofielen, gondwatertrappen en gewassen met het model SWTRE. Rep. 152. Winand Staring Centre, Wageningen, The Netherlands.

Appendix A: Selected Input and Output Files for Hypothetical Example

The test problem illustrates basic and some of the new features of the Farm Process (FMP2). Details of the test problem and results are discussed in the section titled "Example Problem." The entire example input problem is available with the model at the USGS web site indicated in the Preface. The following data sets show the input data sets to help understand the discussion of the example problem structure and selected results that show how the new features allow the more detailed analysis of the hydrologic system. The explanation of files and input data variables are shown in a blue text after the each data entry.

Name File (NAM) Input Data Set

```
list          7      ex1.lst
lpf           9      ex1tr.lpf
bas6         10      ex1tr.ba6
dis          11      ex1tr.dis
ghb          12      ex1tr.ghb
pcg          14      ex1.pcg
oc           15      ex1tr.oc
fmp          16      ex1tr2.fmp
sfr          17      ex1tr.sfr
mnw1         18      ex1_MN.mnw
uzf          19      ex1tr.uzf
hyd          22      hyd_all.hyd
data         21      ex1.hed
data         20      OFE.in
data         30      ROOT.IN
data         40      KC.in
data         50      FTE.in
data         60      INEFFSW.IN
data         80      NRDFAC.IN
data(binary) 70      CBC.OUT
DATA(BINARY) 71      hyd_all.sav
DATA(BINARY) 61      f5&8bin.uzfot    Output file for recharge and ground-water discharge
DATA         62      f5_r2c2.opt1     Output file for cell in row 2, col. 2 of farm 5 using IUZOPT=1
DATA         63      f5_r3c3.opt2     Output file for cell in row 3, col. 3 of farm 5 using IUZOPT=2
DATA         64      f5_r4c4.opt3     Output file for cell in row 4, col. 4 of farm 5 using IUZOPT=3
DATA         65      f5_r5c4.opt2     Output file for cell in row 5, col. 4 of farm 5 using IUZOPT=2
DATA         66      f8_r10c20.opt2   Output file for cell in row 10, col. 20 of farm 8 using IUZOPT=2
DATA         67      f5&8.uzfot       Output file of times series of unsaturated-zone water budgets
```

Basic (BAS) Package Input Data Set

```
# Example model 1
free
constant 1
constant 1
constant 1
constant 1
-999
internal 1. (free) -1
  291.78  291.55  291.26  290.91  290.49  290.00  289.44  288.82  288.14  287.40  286.62  285.82  285.00  284.17  283.33  282.48  281.63  280.76  279.87  278.91
  291.79  291.56  291.28  290.92  290.50  290.00  289.44  288.82  288.13  287.39  286.62  285.81  284.99  284.16  283.32  282.48  281.63  280.78  279.89  278.93
  291.82  291.59  291.31  290.95  290.52  290.02  289.45  288.82  288.13  287.38  286.60  285.80  284.97  284.14  283.31  282.48  281.65  280.82  279.94  278.99
  291.86  291.63  291.35  290.97  290.53  290.02  289.44  288.80  288.10  287.35  286.57  285.77  284.94  284.11  283.29  282.48  281.68  280.88  280.03  279.08
  291.91  291.68  291.40  291.01  290.55  290.03  289.43  288.78  288.07  287.31  286.53  285.73  284.90  284.07  283.27  282.48  281.71  280.93  280.10  279.16
  291.98  291.75  291.46  291.06  290.59  290.04  289.43  288.75  288.03  287.27  286.47  285.67  284.84  284.02  283.23  282.47  281.73  280.96  280.12  279.18
  292.07  291.84  291.54  291.13  290.63  290.06  289.42  288.72  287.97  287.19  286.39  285.58  284.76  283.95  283.17  282.43  281.71  280.94  280.11  279.17
  292.19  291.96  291.65  291.22  290.70  290.09  289.40  288.66  287.88  287.08  286.27  285.45  284.64  283.85  283.08  282.35  281.63  280.87  280.05  279.11
  292.35  292.11  291.80  291.35  290.80  290.14  289.37  288.57  287.75  286.92  286.09  285.29  284.50  283.71  282.95  282.22  281.50  280.73  279.91  278.98
  292.54  292.31  291.99  291.53  290.95  290.23  289.30  288.41  287.55  286.68  285.82  285.04  284.34  283.56  282.78  282.04  281.30  280.52  279.70  278.71
  292.79  292.56  292.24  291.77  291.19  290.47  289.63  288.73  287.82  286.88  285.93  284.97  284.21  283.40  282.59  281.81  281.04  280.24  279.39  278.45
  293.10  292.88  292.56  292.09  291.50  290.78  289.94  289.01  288.03  287.03  286.00  284.93  284.11  283.24  282.37  281.53  280.70  279.86  278.98  278.03
  293.49  293.27  292.95  292.49  291.90  291.17  290.27  289.27  288.23  287.16  286.07  284.96  284.03  283.07  282.11  281.20  280.29  279.37  278.43  277.44
  293.09  292.87  292.54  292.10  291.53  290.81  289.96  289.04  288.07  287.08  286.07  285.07  284.13  283.19  282.27  281.38  280.51  279.70  278.85  277.91
  292.78  292.55  292.23  291.79  291.23  290.52  289.68  288.80  287.89  286.96  286.03  285.09  284.18  283.28  282.39  281.53  280.70  279.95  279.15  278.24
  292.53  292.30  291.98  291.55  291.00  290.29  289.38  288.51  287.66  286.80  285.93  285.05  284.19  283.32  282.47  281.63  280.88  280.14  279.36  278.45
  292.34  292.11  291.79  291.37  290.85  290.20  289.20  288.43  287.82  286.99  286.12  285.25  284.38  283.51  282.66  281.82  281.04  280.28  279.49  278.58
  292.19  291.95  291.64  291.24  290.75  290.14  289.46  288.72  287.94  287.12  286.28  285.41  284.54  283.67  282.81  281.97  281.16  280.36  279.56  278.64
  292.07  291.84  291.53  291.15  290.68  290.11  289.47  288.76  288.01  287.22  286.39  285.53  284.67  283.80  282.93  282.08  281.24  280.39  279.54  278.61
  291.98  291.75  291.45  291.07  290.62  290.08  289.47  288.80  288.06  287.29  286.48  285.63  284.77  283.90  283.03  282.16  281.30  280.42  279.53  278.58
  291.92  291.69  291.39  291.02  290.58  290.05  289.46  288.80  288.07  287.32  286.52  285.69  284.82  283.97  283.10  282.22  281.34  280.45  279.54  278.57
  291.88  291.65  291.36  290.99  290.55  290.04  289.45  288.81  288.10  287.34  286.54  285.73  284.88  284.02  283.14  282.26  281.38  280.48  279.56  278.59
  291.87  291.63  291.34  290.97  290.54  290.03  289.45  288.81  288.11  287.35  286.56  285.74  284.90  284.04  283.17  282.29  281.40  280.50  279.58  278.60
internal 1. (free) -1
  291.37  291.33  291.12  290.79  290.38  289.89  289.34  288.73  288.05  287.33  286.57  285.78  284.98  284.16  283.34  282.53  281.72  280.94  280.22  279.73
  291.39  291.35  291.13  290.80  290.39  289.90  289.34  288.73  288.05  287.32  286.56  285.77  284.97  284.15  283.34  282.52  281.72  280.95  280.24  279.75
  291.41  291.37  291.15  290.82  290.40  289.91  289.35  288.72  288.04  287.31  286.55  285.76  284.95  284.14  283.33  282.52  281.74  280.97  280.28  279.79
  291.45  291.41  291.19  290.85  290.42  289.91  289.34  288.71  288.03  287.29  286.52  285.73  284.92  284.11  283.31  282.52  281.75  281.01  280.33  279.85
  291.50  291.46  291.23  290.88  290.44  289.92  289.34  288.69  287.99  287.25  286.48  285.69  284.88  284.08  283.29  282.52  281.76  281.04  280.36  279.89
  291.57  291.53  291.29  290.93  290.47  289.94  289.33  288.67  287.95  287.20  286.42  285.63  284.82  284.03  283.25  282.49  281.76  281.04  280.37  279.87
  291.66  291.62  291.37  291.00  290.52  289.96  289.33  288.64  287.90  287.13  286.34  285.55  284.74  283.95  283.18  282.44  281.72  281.01  280.35  279.87
  291.77  291.73  291.48  291.09  290.59  290.00  289.32  288.60  287.83  287.04  286.24  285.44  284.64  283.85  283.09  282.36  281.64  280.93  280.27  279.67
  291.92  291.87  291.62  291.21  290.69  290.05  289.33  288.55  287.75  286.93  286.11  285.30  284.51  283.73  282.97  282.23  281.51  280.80  280.14  279.67
  292.09  292.05  291.79  291.37  290.82  290.15  289.36  288.54  287.69  286.84  286.01  285.16  284.38  283.59  282.82  282.07  281.33  280.61  279.94  279.48
  292.30  292.26  292.00  291.57  291.02  290.33  289.54  288.69  287.80  286.90  285.99  285.09  284.27  283.46  282.65  281.87  281.11  280.37  279.69  279.22
  292.51  292.48  292.22  291.79  291.23  290.54  289.74  288.86  287.94  286.99  286.02  285.06  284.19  283.33  282.48  281.66  280.86  280.10  279.40  278.93
  292.67  292.63  292.37  291.95  291.39  290.69  289.88  288.98  288.03  287.05  286.05  285.07  284.15  283.29  282.36  281.50  280.67  279.87  279.16  278.68
  292.51  292.48  292.22  291.80  291.25  290.56  289.76  288.89  287.97  287.02  286.06  285.11  284.18  283.28  282.40  281.54  280.73  279.97  279.28  278.82
  292.29  292.26  292.00  291.59  291.04  290.37  289.58  288.74  287.86  286.96  286.04  285.13  284.23  283.34  282.48  281.65  280.86  280.13  279.47  279.02
  292.09  292.05  291.79  291.39  290.86  290.20  289.42  288.60  287.76  286.90  286.03  285.15  284.28  283.42  282.57  281.76  280.99  280.28  279.63  279.18
  291.91  291.87  291.62  291.23  290.72  290.10  289.38  288.61  287.81  286.98  286.13  285.27  284.41  283.55  282.71  281.90  281.13  280.40  279.75  279.30
  291.77  291.73  291.48  291.11  290.63  290.04  289.37  288.64  287.87  287.07  286.24  285.39  284.53  283.68  282.84  282.02  281.23  280.49  279.83  279.37
```

```
291.66 291.62 291.38 291.02 290.56 290.00 289.37 288.68 287.93 287.15 286.33 285.49 284.64 283.79 282.95 282.12 281.31 280.54 279.85 279.38
 291.58 291.53 291.30 290.95 290.50 289.97 289.37 288.70 287.98 287.21 286.41 285.58 284.73 283.88 283.03 282.19 281.37 280.58 279.87 279.39
 291.52 291.47 291.24 290.90 290.46 289.95 289.36 288.71 288.00 287.24 286.45 285.63 284.80 283.95 283.09 282.25 281.41 280.61 279.88 279.39
 291.48 291.44 291.21 290.87 290.44 289.93 289.36 288.71 288.01 287.27 286.48 285.67 284.84 283.99 283.14 282.29 281.44 280.63 279.90 279.40
 291.46 291.42 291.19 290.85 290.43 289.92 289.35 288.72 288.02 287.28 286.49 285.69 284.86 284.01 283.16 282.31 281.46 280.65 279.91 279.41
internal 1. (free) -1
 291.04 290.95 290.75 290.46 290.08 289.62 289.09 288.50 287.85 287.17 286.45 285.70 284.94 284.19 283.45 282.73 282.07 281.49 281.05 280.79
 291.05 290.96 290.76 290.47 290.08 289.62 289.09 288.50 287.85 287.16 286.44 285.69 284.94 284.18 283.44 282.73 282.07 281.50 281.05 280.80
 291.07 290.98 290.78 290.48 290.10 289.63 289.09 288.49 287.84 287.15 286.42 285.68 284.92 284.17 283.43 282.72 282.07 281.50 281.06 280.81
 291.11 291.02 290.81 290.51 290.11 289.64 289.09 288.49 287.83 287.13 286.40 285.65 284.90 284.14 283.41 282.71 282.06 281.50 281.06 280.82
 291.16 291.06 290.85 290.54 290.14 289.65 289.09 288.48 287.81 287.10 286.37 285.62 284.86 284.11 283.38 282.68 282.05 281.49 281.06 280.81
 291.22 291.12 290.91 290.59 290.17 289.67 289.10 288.47 287.79 287.07 286.33 285.57 284.81 284.06 283.33 282.65 282.01 281.46 281.04 280.79
 291.30 291.20 290.98 290.65 290.22 289.70 289.11 288.46 287.76 287.03 286.27 285.51 284.75 284.00 283.27 282.59 281.96 281.41 280.99 280.75
 291.39 291.29 291.06 290.72 290.27 289.74 289.12 288.45 287.73 286.98 286.21 285.44 284.67 283.92 283.20 282.51 281.88 281.33 280.91 280.67
 291.50 291.40 291.16 290.81 290.35 289.79 289.15 288.45 287.71 286.94 286.15 285.37 284.59 283.83 283.10 282.41 281.78 281.23 280.80 280.56
 291.62 291.51 291.27 290.91 290.43 289.86 289.19 288.45 287.70 286.91 286.10 285.30 284.51 283.74 283.00 282.30 281.66 281.10 280.67 280.43
 291.74 291.63 291.39 291.02 290.53 289.94 289.26 288.51 287.72 286.91 286.07 285.25 284.44 283.65 282.90 282.18 281.53 280.96 280.53 280.28
 291.83 291.72 291.48 291.10 290.61 290.01 289.32 288.56 287.76 286.92 286.07 285.22 284.39 283.58 282.81 282.08 281.41 280.84 280.39 280.15
 291.87 291.76 291.52 291.14 290.65 290.05 289.36 288.59 287.78 286.93 286.07 285.21 284.36 283.54 282.75 282.01 281.34 280.75 280.31 280.06
 291.83 291.72 291.48 291.11 290.62 290.02 289.33 288.58 287.77 286.93 286.07 285.21 284.36 283.54 282.75 282.00 281.33 280.75 280.31 280.06
 291.74 291.63 291.39 291.03 290.54 289.96 289.28 288.53 287.74 286.93 286.07 285.23 284.39 283.57 282.78 282.04 281.37 280.80 280.36 280.11
 291.63 291.52 291.28 290.92 290.45 289.88 289.22 288.49 287.72 286.92 286.09 285.26 284.43 283.62 282.84 282.10 281.44 280.87 280.43 280.19
 291.51 291.40 291.17 290.83 290.37 289.81 289.18 288.47 287.72 286.94 286.13 285.31 284.49 283.69 282.91 282.18 281.51 280.94 280.50 280.26
 291.40 291.30 291.08 290.74 290.30 289.76 289.15 288.47 287.74 286.97 286.18 285.37 284.56 283.76 282.99 282.25 281.58 281.01 280.57 280.32
 291.31 291.21 290.99 290.67 290.24 289.72 289.13 288.47 287.76 287.01 286.23 285.44 284.63 283.84 283.06 282.32 281.65 281.06 280.62 280.37
 291.24 291.14 290.93 290.61 290.19 289.69 289.12 288.48 287.78 287.05 286.28 285.49 284.69 283.90 283.12 282.38 281.70 281.11 280.65 280.40
 291.19 291.09 290.88 290.57 290.16 289.67 289.11 288.48 287.80 287.07 286.31 285.53 284.74 283.95 283.17 282.43 281.74 281.14 280.68 280.42
 291.15 291.05 290.85 290.54 290.14 289.66 289.10 288.48 287.81 287.09 286.34 285.56 284.77 283.98 283.20 282.46 281.77 281.16 280.70 280.44
 291.13 291.04 290.83 290.53 290.13 289.65 289.10 288.48 287.81 287.10 286.35 285.57 284.79 284.00 283.22 282.47 281.78 281.18 280.71 280.45
internal 1. (free) -1
 291.01 290.92 290.72 290.43 290.05 289.59 289.07 288.48 287.84 287.15 286.43 285.69 284.94 284.19 283.45 282.75 282.10 281.54 281.12 280.88
 291.02 290.93 290.73 290.44 290.06 289.60 289.07 288.48 287.83 287.15 286.43 285.69 284.93 284.18 283.45 282.75 282.10 281.54 281.12 280.89
 291.04 290.95 290.75 290.45 290.07 289.60 289.07 288.47 287.82 287.13 286.41 285.67 284.92 284.17 283.44 282.74 282.10 281.55 281.13 280.89
 291.08 290.98 290.78 290.48 290.09 289.61 289.07 288.47 287.81 287.12 286.39 285.65 284.89 284.15 283.42 282.73 282.09 281.54 281.13 280.90
 291.13 291.03 290.82 290.51 290.11 289.63 289.07 288.46 287.79 287.09 286.36 285.61 284.86 284.11 283.39 282.70 282.07 281.53 281.12 280.89
 291.19 291.09 290.87 290.56 290.14 289.65 289.08 288.45 287.77 287.06 286.32 285.56 284.81 284.06 283.34 282.66 282.04 281.50 281.09 280.87
 291.27 291.16 290.94 290.62 290.19 289.67 289.09 288.44 287.75 287.02 286.27 285.51 284.75 284.00 283.28 282.60 281.98 281.45 281.04 280.82
 291.36 291.25 291.03 290.69 290.25 289.70 289.10 288.43 287.72 286.97 286.21 285.44 284.68 283.93 283.21 282.53 281.90 281.37 280.97 280.74
 291.47 291.35 291.12 290.77 290.32 289.76 289.13 288.44 287.70 286.93 286.15 285.37 284.60 283.84 283.11 282.43 281.80 281.27 280.86 280.64
 291.58 291.47 291.23 290.87 290.40 289.83 289.17 288.46 287.70 286.91 286.11 285.31 284.52 283.75 283.01 282.32 281.69 281.15 280.74 280.51
 291.69 291.57 291.33 290.97 290.49 289.89 289.23 288.50 287.72 286.91 286.08 285.26 284.45 283.67 282.92 282.21 281.57 281.02 280.60 280.37
 291.77 291.66 291.41 291.04 290.56 289.97 289.29 288.54 287.74 286.91 286.07 285.23 284.41 283.60 282.84 282.12 281.46 280.90 280.48 280.25
 291.81 291.69 291.45 291.08 290.59 290.00 289.31 288.56 287.76 286.92 286.07 285.22 284.38 283.57 282.79 282.06 281.39 280.83 280.40 280.17
 291.77 291.66 291.42 291.05 290.57 289.98 289.30 288.55 287.75 286.92 286.07 285.22 284.38 283.56 282.78 282.05 281.38 280.82 280.40 280.16
 291.69 291.58 291.34 290.98 290.50 289.92 289.25 288.52 287.73 286.91 286.08 285.24 284.40 283.59 282.81 282.08 281.42 280.86 280.44 280.21
 291.58 291.47 291.24 290.88 290.42 289.85 289.20 288.48 287.71 286.91 286.09 285.27 284.44 283.64 282.86 282.13 281.48 280.92 280.50 280.27
 291.47 291.36 291.13 290.79 290.34 289.79 289.16 288.46 287.71 286.93 286.13 285.32 284.50 283.70 282.93 282.20 281.55 280.99 280.57 280.34
 291.37 291.26 291.04 290.71 290.27 289.74 289.13 288.45 287.73 286.96 286.18 285.37 284.57 283.77 283.00 282.28 281.62 281.06 280.63 280.40
 291.28 291.18 290.96 290.63 290.21 289.70 289.11 288.45 287.75 287.00 286.22 285.43 284.63 283.84 283.07 282.34 281.68 281.11 280.68 280.45
 291.21 291.11 290.90 290.58 290.17 289.67 289.09 288.46 287.76 287.03 286.27 285.48 284.69 283.90 283.13 282.40 281.73 281.16 280.72 280.48
 291.16 291.06 290.85 290.54 290.14 289.65 289.09 288.46 287.78 287.06 286.30 285.52 284.73 283.95 283.18 282.44 281.77 281.19 280.75 280.51
 291.12 291.02 290.82 290.51 290.11 289.63 289.08 288.46 287.79 287.07 286.32 285.55 284.76 283.98 283.21 282.47 281.80 281.21 280.77 280.53
 291.10 291.01 290.80 290.50 290.10 289.63 289.08 288.46 287.79 287.08 286.33 285.56 284.78 284.00 283.22 282.49 281.81 281.22 280.78 280.54
```

Discretization File (DIS) Input Data Set

```
# Example model 1 transient discretization file
4   23  20  24  4  2
1 1 1 0
constant 500
constant 500
OPEN/CLOSE GSE.IN 1.0 (FREE) -1     TOP of layer 1
constant 205                        BOTM of layer 1 / top of confining bed
constant 200                        BOTM of confining bed below layer 1
constant 140                        BOTM of layer 2 / top of confining bed
constant 125                        BOTM of confining bed below layer 2
constant 65                         BOTM of layer 3 / top of confining bed
constant 60                         BOTM of confining bed below layer 3
constant 0                          BOTM of layer 4
30.42 2 1 TR
30.42 2 1 TR
30.42 2 1 TR
30.42 2 1 TR
30.42 2 1 TR
30.42 2 1 TR
30.42 2 1 TR
30.42 2 1 TR
30.42 2 1 TR
30.42 2 1 TR
30.42 2 1 TR
30.42 2 1 TR
30.42 2 1 TR
30.42 2 1 TR
30.42 2 1 TR
30.42 2 1 TR
30.42 2 1 TR
30.42 2 1 TR
30.42 2 1 TR
30.42 2 1 TR
30.42 2 1 TR
30.42 2 1 TR
30.42 2 1 TR
30.42 2 1 TR
```

Layer Property Flow (LPF) Input Data Set

```
70  2222  0        ILPFCB HDRY NPLPF
1  0  0  0         LAYTYP (UNCONFINED (non-0: in LPF automatically convertible) / 3 CONFINED LAYERS (0))
0  0  0  0         LAYAVG (HARMONIC MEAN)
1. 1. 1. 1.        CHANI (HORIZONTAL ANISOTROPY)
1  1  1  1         LAYVKA (VKA IS VERTICAL HYDRAULIC CONDUCTIVITY)
0  0  0  0         LAYWET (WETTING INACTIVE)
CONSTANT 10.       HK OF LAYER 1
CONSTANT 1         VKA RATIO VERTICAL TO HORIZONTAL HK OF LAYER 1
CONSTANT 0.00001   SS OF LAYER 1
CONSTANT 0.02      SY SPECIFIC YIELD IF TRANSIENT/UNCONFINED
CONSTANT 0.002     VKCB OF CONFINING BED BELOW LAYER 1
CONSTANT 1.5       HK OF LAYER 2
CONSTANT 1         VKA RATIO VERTICAL TO HORIZONTAL HK OF LAYER 2
CONSTANT 0.00001   SS OF LAYER 2
CONSTANT 0.001     VKCB OF CONFINING BED BELOW LAYER 2
CONSTANT 1.0       HK OF LAYER 3
CONSTANT 1         VKA RATIO VERTICAL TO HORIZONTAL HK OF LAYER 3
CONSTANT 0.00001   SS OF LAYER 3
CONSTANT 0.0005    VKCB OF CONFINING BED BELOW LAYER 3
CONSTANT 0.15      HK OF LAYER 4
CONSTANT 1         VKA RATIO VERTICAL TO HORIZONTAL HK OF LAYER 4
CONSTANT 0.00001   SS OF LAYER 4
```

Preconditioned Conjugate Gradient (PCG) Package Input Data Set

```
100 10 1            NPCOND 1 - mod. incomplete, 2 - polynomial
0.001 10 1. 2 1 1 1 HCLOSE(head),RCLOSE(FLOWRATE: should be rel. high for m3/day)
```

Output Control (OC) Package Input Data Set

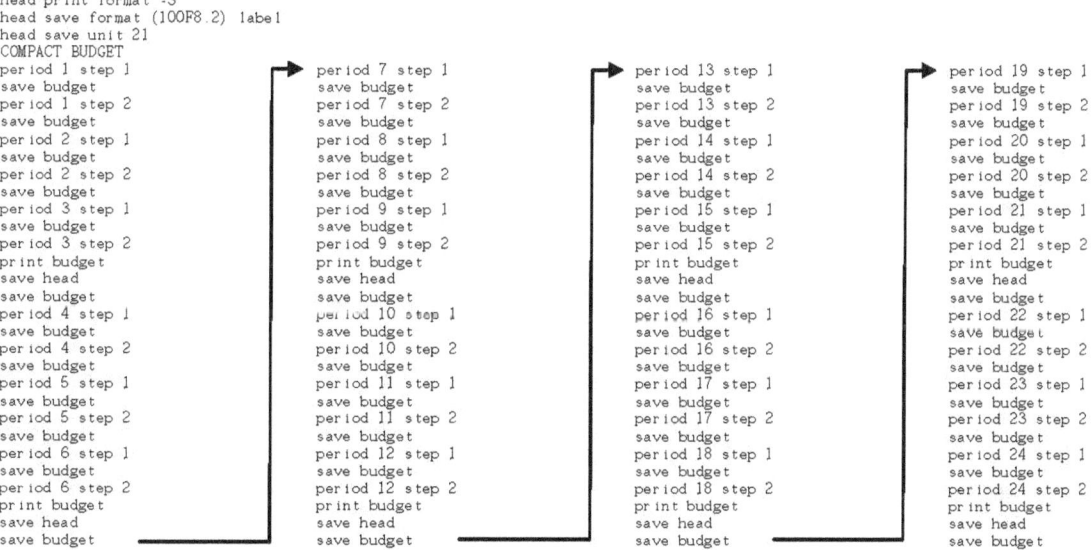

```
head print format -3
head save format (100F8.2) label
head save unit 21
COMPACT BUDGET
period 1 step 1        period 7 step 1        period 13 step 1       period 19 step 1
save budget            save budget            save budget            save budget
period 1 step 2        period 7 step 2        period 13 step 2       period 19 step 2
save budget            save budget            save budget            save budget
period 2 step 1        period 8 step 1        period 14 step 1       period 20 step 1
save budget            save budget            save budget            save budget
period 2 step 2        period 8 step 2        period 14 step 2       period 20 step 2
save budget            save budget            save budget            save budget
period 3 step 1        period 9 step 1        period 15 step 1       period 21 step 1
save budget            save budget            save budget            save budget
period 3 step 2        period 9 step 2        period 15 step 2       period 21 step 2
print budget          print budget           print budget           print budget
save head              save head              save head              save head
save budget            save budget            save budget            save budget
period 4 step 1        period 10 step 1       period 16 step 1       period 22 step 1
save budget            save budget            save budget            save budget
period 4 step 2        period 10 step 2       period 16 step 2       period 22 step 2
save budget            save budget            save budget            save budget
period 5 step 1        period 11 step 1       period 17 step 1       period 23 step 1
save budget            save budget            save budget            save budget
period 5 step 2        period 11 step 2       period 17 step 2       period 23 step 2
save budget            save budget            save budget            save budget
period 6 step 1        period 12 step 1       period 18 step 1       period 24 step 1
save budget            save budget            save budget            save budget
period 6 step 2        period 12 step 2       period 18 step 2       period 24 step 2
print budget          print budget           print budget           print budget
save head              save head              save head              save head
save budget            save budget            save budget            save budget
```

General Head Boundary (GHB) Package Input Data Set

```
92  70
92  0
1   1    1    296.17    39
1   2    1    296.17    39
1   3    1    296.17    39
1   4    1    296.17    39
1   5    1    296.17    39
1   6    1    296.17    39
1   7    1    296.17    39
1   8    1    296.17    39
1   9    1    296.17    39
1   10   1    296.17    39
1   11   1    296.17    39
1   12   1    296.17    39
1   13   1    296.17    39
1   14   1    296.17    39
1   15   1    296.17    39
1   16   1    296.17    39
1   17   1    296.17    39
1   18   1    296.17    39
1   19   1    296.17    39
1   20   1    296.17    39
1   21   1    296.17    39
```

```
1    22     1     296.17          39
1    23     1     296.17          39
1    1      20    253.9845581     32.415
1    2      20    253.9845581     32.415
1    3      20    253.9845581     32.415
1    4      20    253.9845581     32.415
1    5      20    253.9845581     32.415
1    6      20    253.9845581     32.415
1    7      20    253.9845581     32.415
1    8      20    253.9845581     32.415
1    9      20    253.9845581     32.415
1    10     20    253.9845581     32.415
1    11     20    253.9845581     32.415
1    12     20    253.9845581     32.415
1    13     20    253.9845581     32.415
1    14     20    253.9845581     32.415
1    15     20    253.9845581     32.415
1    16     20    253.9845581     32.415
1    17     20    253.9845581     32.415
1    18     20    253.9845581     32.415
1    19     20    253.9845581     32.415
1    20     20    253.9845581     32.415
1    21     20    253.9845581     32.415
1    22     20    253.9845581     32.415
1    23     20    253.9845581     32.415
2    1      1     284.17          4.5
2    2      1     284.17          4.5
2    3      1     284.17          4.5
2    4      1     284.17          4.5
2    5      1     284.17          4.5
2    6      1     284.17          4.5
2    7      1     284.17          4.5
2    8      1     284.17          4.5
2    9      1     284.17          4.5
2    10     1     284.17          4.5
2    11     1     284.17          4.5
2    12     1     284.17          4.5
2    13     1     284.17          4.5
2    14     1     284.17          4.5
2    15     1     284.17          4.5
2    16     1     284.17          4.5
2    17     1     284.17          4.5
2    18     1     284.17          4.5
2    19     1     284.17          4.5
2    20     1     284.17          4.5
2    21     1     284.17          4.5
2    22     1     284.17          4.5
2    23     1     284.17          4.5
2    1      20    283.66          4.5
2    2      20    283.66          4.5
2    3      20    283.66          4.5
2    4      20    283.66          4.5
2    5      20    283.66          4.5
2    6      20    283.66          4.5
2    7      20    283.66          4.5
2    8      20    283.66          4.5
2    9      20    283.66          4.5
2    10     20    283.66          4.5
2    11     20    283.66          4.5
2    12     20    283.66          4.5
2    13     20    283.66          4.5
2    14     20    283.66          4.5
2    15     20    283.66          4.5
2    16     20    283.66          4.5
2    17     20    283.66          4.5
2    18     20    283.66          4.5
2    19     20    283.66          4.5
2    20     20    283.66          4.5
2    21     20    283.66          4.5
2    22     20    283.66          4.5
2    23     20    283.66          4.5
-1   0
-1   0
-1   0
-1   0
-1   0
-1   0
-1   0
-1   0
-1   0
-1   0
-1   0
-1   0
-1   0
-1   0
-1   0
-1   0
-1   0
-1   0
-1   0
-1   0
-1   0
-1   0
-1   0
-1   0
```

Streamflow Routing (SFR) Package Input Data Set

```
# Example model 1 - New Stream Aquifer Packagee
100 11 0 0 86400 0.0001 70 0
1   13    1      1      1    500
1   13    2      1      2    500
1   13    3      1      3    250
1   13    3      2      1    250
1   12    3      2      2    500
1   11    3      2      3    500
1   10    3      2      4    500
1   9     3      2      5    500
1   8     3      2      6    500
1   7     3      2      7    500
1   6     3      2      8    500
1   5     3      2      9    500
1   4     3      2     10    500
1   3     3      2     11    500
1   3     4      2     12    500
1   3     5      2     13    500
1   3     6      2     14    500
1   3     7      2     15    500
1   3     8      2     16    500
1   3     9      2     17    500
1   3    10      2     18    500
1   3    11      2     19    500
1   3    12      2     20    500
1   4    12      2     21    500
1   5    12      2     22    500
1   6    12      2     23    500
1   7    12      2     24    500
1   8    12      2     25    500
1   9    12      2     26    500
1   10   12      2     27    250
1   10    7      3      1    250
1   10    8      3      2    500
1   10    9      3      3    500
1   10   10      3      4    500
1   10   11      3      5    500
1   10   12      3      6    250
1   10   12      4      1    250
1   11   12      4      2    500
1   12   12      4      3    500
1   13   12      4      4    250
1   13    3      5      1    250
1   13    4      5      2    500
1   13    5      5      3    250
1   13    5      6      1    250
1   14    5      6      2    500
1   15    5      6      3    500
1   16    5      6      4    500
1   17    5      6      5    500
1   18    5      6      6    500
1   19    5      6      7    500
1   20    5      6      8    500
1   20    6      6      9    500
1   20    7      6     10    500
1   20    8      6     11    500
1   20    9      6     12    500
1   20   10      6     13    500
1   20   11      6     14    500
1   20   12      6     15    500
1   20   13      6     16    500
1   20   14      6     17    500
1   20   15      6     18    500
1   20   16      6     19    500
1   20   17      6     20    500
1   19   17      6     21    500
1   18   17      6     22    500
1   17   17      6     23    500
1   16   17      6     24    250
1   16    7      7      1    250
1   16    8      7      2    500
1   16    9      7      3    500
1   16   10      7      4    500
1   16   11      7      5    500
1   16   12      7      6    500
1   16   13      7      7    500
1   16   14      7      8    500
1   16   15      7      9    500
1   16   16      7     10    500
1   16   17      7     11    250
1   16   17      8      1    250
1   15   17      8      2    500
1   14   17      8      3    500
1   13   17      8      4    250
1   13    5      9      1    250
1   13    6      9      2    500
1   13    7      9      3    500
1   13    8      9      4    500
1   13    9      9      5    500
1   13   10      9      6    500
1   13   11      9      7    500
```

NSTRM NSS NSTRPAR NPARSEG CONST DLEAK ISTCB1 ISTCB2
KRCH IRCH JRCH ISTRM4 ISTRM5 RCHLEN

```
1    13        12        9         8        250
1    13        12        10        1        250
1    13        13        10        2        500
1    13        14        10        3        500
1    13        15        10        4        500
1    13        16        10        5        500
1    13        17        10        6        250
1    13        17        11        1        250
1    13        18        11        2        500
1    13        19        11        3        500
1    13        20        11        4        500
11   2  0                                              ITMP IRDFLG(2= slope of canals follows ground-surface at interpolated depth) IPTFLG(0 results printed!)
1 1 5 0 100000. 0. 0. 0. 0.04                          STREAM:
0.2 1. 298.50 6.                                       GSE at boundary = 299.00; starting elevation = GSE-0.5m          HCOND1 THICKM1 ELEVUP WIDTH1 (DEPTH1)
0.2 1. 295.80 6.                                       GSE at node = 296.55; end elevation = GSE-0.75m                  HCOND2 THICKM2 ELEVDN WIDTH2 (DEPTH2)
2 1 4 1 0 10000. 0. 0. 0. 0.03                         DIVERSION CANAL:
0.01 1. 296.55 3.                                      GSE at node = 296.55; starting elevation = GSE (Canals sits on GSE)
0.01 1. 286.88 3.                                      GSE at node = 286.88; end elevation = GSE (Canal sits on GSE)-> equal distance (0) from start to end from GSE
3 1 4 0 0. 0. 0. 0. 0.03                               TRIBUTARTY DRAIN:
1. 1. 288.90 3.                                        GSE at node = 292.40; starting elevation = GSE-3.5m
1. 1. 284.58 3.                                        GSE at node = 286.88: end elevation = GSE-2.3m = starting elevation of seg.4
4 1 10 0 0. 0. 0. 0. 0.03                              BYPASS CANAL BACK INTO STREAM:
1. 1. 284.58 3.                                        GSE at node = 286.88; starting elevation = GSE-2.3m
1. 1. 284.28 3.                                        GSE at node = 286.58; end elevation = GSE-2.3m
5 1 9 0 0. 0. 0. 0. 0.04                               STREAM:
0.2 1. 295.80 6.                                       GSE at node = 296.55; starting elevation = GSE-0.75m
0.2 1. 293.61 6.                                       GSE at node = 294.61; end elevation = GSE-1m
6 1 8 5 0 10000. 0. 0. 0. 0.03                         DIVERSION CANAL:
0.01 1. 294.61 3.                                      GSE at node = 294.61; starting elevation = GSE (Canals sits on GSE)
0.01 1. 281.47 3.                                      GSE at node = 281.47; end elevation = GSE (Canal sits on GSE)-> equal distance (0) from start to end from GSE
7 1 8 0 0. 0. 0. 0. 0.03                               TRIBUTARTY DRAIN:
1. 1. 289.02 3.                                        GSE at node = 291.32; starting elevation = GSE-2.3m
1. 1. 280.57 3.                                        GSE at node = 281.47; end elevation = GSE-0.9m = starting elevation of seg.8
8 1 8 0 0. 0. 0. 0. 0.03                               BYPASS CANAL BACK INTO STREAM:
1. 1. 280.57 3.                                        GSE at node = 281.47; starting elevation = GSE-0.9m
1. 1. 279.94 3.                                        GSE at node = 280.84; end elevation = GSE-0.9m
9 1 10 0 0. 0. 0. 0. 0.04                              STREAM:
0.2 1. 293.61 6.                                       GSE at node = 294.61; starting elevation = GSE-1m
0.2 1. 284.58 6.                                       GSE at node = 286.58; end elevation = GSE-2m
10 1 11 0 0. 0. 0. 0. 0.04                             STREAM:
0.2 1. 284.58 6.                                       GSE at node = 286.58; starting elevation = GSE-2m
0.2 1. 278.84 6.                                       GSE at node = 280.84; end elevation = GSE-2m
11 1 0 0 0. 0. 0. 0. 0.04                              STREAM:
0.2 1. 278.84 6.                                       GSE at node = 280.84; starting elevation = GSE-2m
0.2 1. 274.83 6.                                       GSE at boundary = 276.83; end elevation = GSE-2m
-1 2 0
-1 2 0
-1 2 0
-1 2 0
-1 2 0
-1 2 0
-1 2 0
-1 2 0
-1 2 0
-1 2 0
-1 2 0
11   2  0                                              ITMP IRDFLG(2= slope of canals follows ground-surface at interpolated depth) IPTFLG(0 results printed!)
1 1 5 0 50000. 0. 0. 0. 0.04                           STREAM:
0.2 1. 298.50 6.                                       GSE at boundary = 299.00; starting elevation = GSE-0.5m          HCOND1 THICKM1 ELEVUP WIDTH1 (DEPTH1)
0.2 1. 295.80 6.                                       GSE at node = 296.55; end elevation = GSE-0.75m                  HCOND2 THICKM2 ELEVDN WIDTH2 (DEPTH2)
2 1 4 1 0 8000. 0. 0. 0. 0.03                          DIVERSION CANAL:
0.01 1. 296.55 3.                                      GSE at node = 296.55; starting elevation = GSE (Canals sits on GSE)
0.01 1. 286.88 3.                                      GSE at node = 286.88; end elevation = GSE (Canal sits on GSE)-> equal distance (0) from start to end from GSE
3 1 4 0 0. 0. 0. 0. 0.03                               TRIBUTARTY DRAIN:
1. 1. 288.90 3.                                        GSE at node = 292.40; starting elevation = GSE-3.5m
1. 1. 284.58 3.                                        GSE at node = 286.88: end elevation = GSE-2.3m = starting elevation of seg.4
4 1 10 0 0. 0. 0. 0. 0.03                              BYPASS CANAL BACK INTO STREAM:
1. 1. 284.58 3.                                        GSE at node = 286.88; starting elevation = GSE-2.3m
1. 1. 284.28 3.                                        GSE at node = 286.58; end elevation = GSE-2.3m
5 1 9 0 0. 0. 0. 0. 0.04                               STREAM:
0.2 1. 295.80 6.                                       GSE at node = 296.55; starting elevation = GSE-0.75m
0.2 1. 293.61 6.                                       GSE at node = 294.61; end elevation = GSE-1m
6 1 8 5 0 10000. 0. 0. 0. 0.03                         DIVERSION CANAL:
0.01 1. 294.61 3.                                      GSE at node = 294.61; starting elevation = GSE (Canals sits on GSE)
0.01 1. 281.47 3.                                      GSE at node = 281.47; end elevation = GSE (Canal sits on GSE)-> equal distance (0) from start to end from GSE
7 1 8 0 0. 0. 0. 0. 0.03                               TRIBUTARTY DRAIN:
1. 1. 289.02 3.                                        GSE at node = 291.32; starting elevation = GSE-2.3m
1. 1. 280.57 3.                                        GSE at node = 281.47; end elevation = GSE-0.9m = starting elevation of seg.8
8 1 8 0 0. 0. 0. 0. 0.03                               BYPASS CANAL BACK INTO STREAM:
1. 1. 280.57 3.                                        GSE at node = 281.47; starting elevation = GSE-0.9m
1. 1. 279.94 3.                                        GSE at node = 280.84; end elevation = GSE-0.9m
9 1 10 0 0. 0. 0. 0. 0.04                              STREAM:
0.2 1. 293.61 6.                                       GSE at node = 294.61; starting elevation = GSE-1m
0.2 1. 284.58 6.                                       GSE at node = 286.58; end elevation = GSE-2m
10 1 11 0 0. 0. 0. 0. 0.04                             STREAM:
0.2 1. 284.58 6.                                       GSE at node = 286.58; starting elevation = GSE-2m
0.2 1. 278.84 6.                                       GSE at node = 280.84; end elevation = GSE-2m
11 1 0 0 0. 0. 0. 0. 0.04                              STREAM:
0.2 1. 278.84 6.                                       GSE at node = 280.84; starting elevation = GSE-2m
0.2 1. 274.83 6.                                       GSE at boundary = 276.83; end elevation = GSE-2m
-1 2 0
-1 2 0
-1 2 0
-1 2 0
-1 2 0
```

```
-1 2 0
-1 2 0
-1 2 0
-1 2 0
-1 2 0
-1 2 0
```

Unsaturated Zone Flow (UZF) Package Input Data Sets

```
3 1 1 0 61 70 25 20 6 0.2
INTERNAL 1 (FREE) -1
  0  0  0  0  0  0  0  0  0  0  0  0  0  0  0  0  0  0  0  0
  0  1  1  1  0  0  0  0  0  0  0  0  0  0  0  0  0  0  0  0
  0  1  1  1  0  0  0  0  0  0  0  0  0  0  0  0  0  0  0  0
  0  1  1  1  0  0  0  0  0  0  0  0  0  0  0  0  0  0  0  0
  0  1  1  1  0  0  0  0  0  0  0  0  0  0  0  0  0  0  0  0
  0  0  0  0  0  0  0  0  0  0  0  0  0  0  0  0  0  0  0  0
  0  0  0  0  0  0  0  0  0  0  0  0  0  0  0  0  0  0  0  0
  0  0  0  0  0  0  0  0  0  0  0  0  0  0  0  0  0  0  1  1
  0  0  0  0  0  0  0  0  0  0  0  0  0  0  0  0  0  0  1  1
  0  0  0  0  0  0  0  0  0  0  0  0  0  0  0  0  0  0  1  1
  0  0  0  0  0  0  0  0  0  0  0  0  0  0  0  0  0  0  1  1
  0  0  0  0  0  0  0  0  0  0  0  0  0  0  0  1  1  1  1  1
  0  0  0  0  0  0  0  0  0  0  0  0  0  0  0  0  0  0  1  1
  0  0  0  0  0  0  0  0  0  0  0  0  0  0  0  0  0  0  1  1
  0  0  0  0  0  0  0  0  0  0  0  0  0  0  0  0  0  0  1  1
  0  0  0  0  0  0  0  0  0  0  0  0  0  0  0  0  0  0  1  1
  0  0  0  0  0  0  0  0  0  0  0  0  0  0  0  0  0  0  0  0
  0  0  0  0  0  0  0  0  0  0  0  0  0  0  0  0  0  0  0  0
  0  0  0  0  0  0  0  0  0  0  0  0  0  0  0  0  0  0  0  0
  0  0  0  0  0  0  0  0  0  0  0  0  0  0  0  0  0  0  0  0
  0  0  0  0  0  0  0  0  0  0  0  0  0  0  0  0  0  0  0  0
INTERNAL 1 (FREE) -1
  0  0  0  0  0  0  0  0  0  0  0  0  0  0  0  0  0  0  0  0
  0  2  2  2  0  0  0  0  0  0  0  0  0  0  0  0  0  0  0  0
  0  2  2  2  0  0  0  0  0  0  0  0  0  0  0  0  0  0  0  0
  0  2  2  2  0  0  0  0  0  0  0  0  0  0  0  0  0  0  0  0
  0  2  2  2  0  0  0  0  0  0  0  0  0  0  0  0  0  0  0  0
  0  0  0  0  0  0  0  0  0  0  0  0  0  0  0  0  0  0  0  0
  0  0  0  0  0  0  0  0  0  0  0  0  0  0  0  0  0  0  0  0
  0  0  0  0  0  0  0  0  0  0  0  0  0  0  0  0  0 11 11
  0  0  0  0  0  0  0  0  0  0  0  0  0  0  0  0  0 11 11
  0  0  0  0  0  0  0  0  0  0  0  0  0  0  0  0  0 11 11
  0  0  0  0  0  0  0  0  0  0  0  0  0  0  0  0  0 11 11
  0  0  0  0  0  0  0  0  0  0  0  0  0  0 11 11 11 11 11
  0  0  0  0  0  0  0  0  0  0  0  0  0  0  0  0  0 11 11
  0  0  0  0  0  0  0  0  0  0  0  0  0  0  0  0  0 11 11
  0  0  0  0  0  0  0  0  0  0  0  0  0  0  0  0  0 11 11
  0  0  0  0  0  0  0  0  0  0  0  0  0  0  0  0  0 11 11
  0  0  0  0  0  0  0  0  0  0  0  0  0  0  0  0  0  0  0
  0  0  0  0  0  0  0  0  0  0  0  0  0  0  0  0  0  0  0
  0  0  0  0  0  0  0  0  0  0  0  0  0  0  0  0  0  0  0
  0  0  0  0  0  0  0  0  0  0  0  0  0  0  0  0  0  0  0
  0  0  0  0  0  0  0  0  0  0  0  0  0  0  0  0  0  0  0
INTERNAL 1.E-3 (FREE) -1
 0. 0. 0. 0. 0. 0. 0. 0. 0. 0. 0. 0. 0. 0. 0. 0. 0. 0. 0. 0.
 0. 1. 1. 1. 0. 0. 0. 0. 0. 0. 0. 0. 0. 0. 0. 0. 0. 0. 0. 0.
 0. 1. 1. 1. 0. 0. 0. 0. 0. 0. 0. 0. 0. 0. 0. 0. 0. 0. 0. 0.
 0. 1. 1. 1. 0. 0. 0. 0. 0. 0. 0. 0. 0. 0. 0. 0. 0. 0. 0. 0.
 0. 1. 1. 1. 0. 0. 0. 0. 0. 0. 0. 0. 0. 0. 0. 0. 0. 0. 0. 0.
 0. 0. 0. 0. 0. 0. 0. 0. 0. 0. 0. 0. 0. 0. 0. 0. 0. 0. 0. 0.
 0. 0. 0. 0. 0. 0. 0. 0. 0. 0. 0. 0. 0. 0. 0. 0. 0. 0. 0. 0.
 0. 0. 0. 0. 0. 0. 0. 0. 0. 0. 0. 0. 0. 0. 0. 0. 0. 0. 1. 1.
 0. 0. 0. 0. 0. 0. 0. 0. 0. 0. 0. 0. 0. 0. 0. 0. 0. 0. 1. 1.
 0. 0. 0. 0. 0. 0. 0. 0. 0. 0. 0. 0. 0. 0. 0. 0. 0. 0. 1. 1.
 0. 0. 0. 0. 0. 0. 0. 0. 0. 0. 0. 0. 0. 0. 0. 0. 0. 0. 1. 1.
 0. 0. 0. 0. 0. 0. 0. 0. 0. 0. 0. 0. 0. 0. 1. 1. 1. 1. 1.
 0. 0. 0. 0. 0. 0. 0. 0. 0. 0. 0. 0. 0. 0. 0. 0. 0. 0. 1. 1.
 0. 0. 0. 0. 0. 0. 0. 0. 0. 0. 0. 0. 0. 0. 0. 0. 0. 0. 1. 1.
 0. 0. 0. 0. 0. 0. 0. 0. 0. 0. 0. 0. 0. 0. 0. 0. 0. 0. 1. 1.
 0. 0. 0. 0. 0. 0. 0. 0. 0. 0. 0. 0. 0. 0. 0. 0. 0. 0. 1. 1.
 0. 0. 0. 0. 0. 0. 0. 0. 0. 0. 0. 0. 0. 0. 0. 0. 0. 0. 0. 0.
 0. 0. 0. 0. 0. 0. 0. 0. 0. 0. 0. 0. 0. 0. 0. 0. 0. 0. 0. 0.
 0. 0. 0. 0. 0. 0. 0. 0. 0. 0. 0. 0. 0. 0. 0. 0. 0. 0. 0. 0.
 0. 0. 0. 0. 0. 0. 0. 0. 0. 0. 0. 0. 0. 0. 0. 0. 0. 0. 0. 0.
 0. 0. 0. 0. 0. 0. 0. 0. 0. 0. 0. 0. 0. 0. 0. 0. 0. 0. 0. 0.
CONSTANT 3.5
CONSTANT 0.2
CONSTANT 0.16
 2 2 62 1
 3 3 63 2
 4 4 64 1
 5 4 65 2
 10 20 66 2
-67
1
INTERNAL 1.E-5 (FREE) 0
 0. 0. 0. 0. 0. 0. 0. 0. 0. 0. 0. 0. 0. 0. 0. 0. 0. 0. 0. 0.
 0. 1. 1. 1. 0. 0. 0. 0. 0. 0. 0. 0. 0. 0. 0. 0. 0. 0. 0. 0.
```

```
0. 1. 1. 1. 0. 0. 0. 0. 0. 0. 0. 0. 0. 0. 0. 0. 0. 0. 0. 0.
0. 1. 1. 1. 0. 0. 0. 0. 0. 0. 0. 0. 0. 0. 0. 0. 0. 0. 0. 0.
0. 1. 1. 1. 0. 0. 0. 0. 0. 0. 0. 0. 0. 0. 0. 0. 0. 0. 0. 0.
0. 0. 0. 0. 0. 0. 0. 0. 0. 0. 0. 0. 0. 0. 0. 0. 0. 0. 0. 0.
0. 0. 0. 0. 0. 0. 0. 0. 0. 0. 0. 0. 0. 0. 0. 0. 0. 0. 0. 0.
0. 0. 0. 0. 0. 0. 0. 0. 0. 0. 0. 0. 0. 0. 0. 0. 0. 0. 1. 1.
0. 0. 0. 0. 0. 0. 0. 0. 0. 0. 0. 0. 0. 0. 0. 0. 0. 0. 1. 1.
0. 0. 0. 0. 0. 0. 0. 0. 0. 0. 0. 0. 0. 0. 0. 0. 0. 0. 1. 1.
0. 0. 0. 0. 0. 0. 0. 0. 0. 0. 0. 0. 0. 0. 0. 0. 0. 0. 1. 1.
0. 0. 0. 0. 0. 0. 0. 0. 0. 0. 0. 0. 0. 0. 0. 1. 1. 1. 1. 1.
0. 0. 0. 0. 0. 0. 0. 0. 0. 0. 0. 0. 0. 0. 0. 0. 0. 0. 1. 1.
0. 0. 0. 0. 0. 0. 0. 0. 0. 0. 0. 0. 0. 0. 0. 0. 0. 0. 1. 1.
0. 0. 0. 0. 0. 0. 0. 0. 0. 0. 0. 0. 0. 0. 0. 0. 0. 0. 1. 1.
0. 0. 0. 0. 0. 0. 0. 0. 0. 0. 0. 0. 0. 0. 0. 0. 0. 0. 1. 1.
0. 0. 0. 0. 0. 0. 0. 0. 0. 0. 0. 0. 0. 0. 0. 0. 0. 0. 0. 0.
0. 0. 0. 0. 0. 0. 0. 0. 0. 0. 0. 0. 0. 0. 0. 0. 0. 0. 0. 0.
0. 0. 0. 0. 0. 0. 0. 0. 0. 0. 0. 0. 0. 0. 0. 0. 0. 0. 0. 0.
0. 0. 0. 0. 0. 0. 0. 0. 0. 0. 0. 0. 0. 0. 0. 0. 0. 0. 0. 0.
-1
-1
-1
-1
-1
-1
-1
-1
-1
-1
-1
-1
-1
-1
-1
-1
-1
-1
-1
-1
-1
-1
```

Multi-Node Well (MNW) Input Data Sets

```
        18        70         0
SKIN
FILE:t.wl1          WEL1:91
FILE:t.ByNode       BYNODE:92   ALLTIME
FILE:t.Qsum         QSUM:93     ALLTIME
        18
         1    6    2    0.      -1 0.1 1 205    1.E16
         2    6    2    0. MN -1 0.1 1 205    1.E16
         3    6    2    0. MN -1 0.1 1 205    1.E16
         4    6    2    0. MN -1 0.1 1 205    1.E16
         1    8    2    0.      -1 0.1 1 205    1.E16
         2    8    2    0. MN -1 0.1 1 205    1.E16
         3    8    2    0. MN -1 0.1 1 205    1.E16
         4    8    2    0. MN -1 0.1 1 205    1.E16
         1   10    2    0.      -1 0.1 1 205    1.E16
         2   10    2    0. MN -1 0.1 1 205    1.E16
         3   10    2    0. MN -1 0.1 1 205    1.E16
         4   10    2    0. MN -1 0.1 1 205    1.E16
         2    3   19    0.      -1 0.1 1 205    1.E16
         3    3   19    0. MN -1 0.1 1 205    1.E16
         4    3   19    0. MN -1 0.1 1 205    1.E16
         2    6   17    0.      -1 0.1 1 205    1.E16
         3    6   17    0. MN -1 0.1 1 205    1.E16
         4    6   17    0. MN -1 0.1 1 205    1.E16
-1
-1
-1
-1
-1
-1
-1
-1
-1
-1
-1
-1
-1
-1
-1
-1
-1
-1
-1
-1
-1
-1
```

Farm Process (FMP) Input Data Sets

```
# FMP2 Example model -- ZERO SCENARIO / EQUAL APPROPRATION
PARAMETER 1 15
FLAG_BLOCKS
15 8 6 3              Dimensions (NFARMS NCROPS NSOILS)
2 -1 2 2 2 2         When-to-read Flags (IRTFL ICUFL IPFL IFTEFL IIESWFL IEFFL)
1 0 0                Water Policy Flags (IEBFL IROTFL IDEFFL)
3                    Consumptive Use Concept Flag (ICCFL)
1 1 1 0   1 -1  1    Surface-Water Flags (INRDFL MXNRDT ISRDFL IRDFL ISRRFL IRRFL IALLOT)
70 70 1 2 2          Print Flags or Units (IFWLCB IFNRCB ISDPFL IFBPFL IRTPFL)
AUX QMAXRESET        Flags for Auxiliary Variables
NOOPT                Flags for Options
WELLS1 QMAX 1000.0 15
OPEN/CLOSE WELLS.IN FREE -1
OPEN/CLOSE GSE.IN 1.0 (FREE) -1
OPEN/CLOSE FID.IN 1 (FREE) -1
OPEN/CLOSE SID.IN 1 (FREE) -1
OPEN/CLOSE SOILLIST.IN
OPEN/CLOSE CID.IN 1 (FREE) -1
OPEN/CLOSE PSI.IN
OPEN/CLOSE SRD.IN
OPEN/CLOSE SRR.IN
```

`0 1   SP 1`	`0 1   SP 7`	`0 1   SP 13`	`0 1   SP 19`
`WELLS1`	`WELLS1`	`WELLS1`	`WELLS1`
`EXTERNAL 20 OFE.IN`	`EXTERNAL 20 OFE.IN`	`EXTERNAL 20 OFE.IN`	`EXTERNAL 20 OFE.IN`
`EXTERNAL 30 ROOT.IN`	`EXTERNAL 30 ROOT.IN`	`EXTERNAL 30 ROOT.IN`	`EXTERNAL 30 ROOT.IN`
`EXTERNAL 40 KC.IN`	`EXTERNAL 40 KC.IN`	`EXTERNAL 40 KC.IN`	`EXTERNAL 40 KC.IN`
`CONSTANT 0.00091`	`CONSTANT 0.00692`	`CONSTANT 0.00091`	`CONSTANT 0.00692`
`EXTERNAL 50 FTE.IN`	`EXTERNAL 50 FTE.IN`	`EXTERNAL 50 FTE.IN`	`EXTERNAL 50 FTE.IN`
`EXTERNAL 60 INEFFSW.IN`	`EXTERNAL 60 INEFFSW.IN`	`EXTERNAL 60 INEFFSW.IN`	`EXTERNAL 60 INEFFSW.IN`
`CONSTANT 0.00207`	`CONSTANT 0.00003`	`CONSTANT 0.00207`	`CONSTANT 0.00003`
`OPEN/CLOSE NRDV.IN`	`OPEN/CLOSE NRDV.IN`	`OPEN/CLOSE NRDV.IN`	`OPEN/CLOSE NRDV.IN`
`0`	`0.2`	`0`	`0.05`
`0 1   SP 2`	`0 1   SP 8`	`0 1   SP 14`	`0 1   SP 20`
`WELLS1`	`WELLS1`	`WELLS1`	`WELLS1`
`EXTERNAL 20 OFE.IN`	`EXTERNAL 20 OFE.IN`	`EXTERNAL 20 OFE.IN`	`EXTERNAL 20 OFE.IN`
`EXTERNAL 30 ROOT.IN`	`EXTERNAL 30 ROOT.IN`	`EXTERNAL 30 ROOT.IN`	`EXTERNAL 30 ROOT.IN`
`EXTERNAL 40 KC.IN`	`EXTERNAL 40 KC.IN`	`EXTERNAL 40 KC.IN`	`EXTERNAL 40 KC.IN`
`CONSTANT 0.0015`	`CONSTANT 0.00617`	`CONSTANT 0.0015`	`CONSTANT 0.00617`
`EXTERNAL 50 FTE.IN`	`EXTERNAL 50 FTE.IN`	`EXTERNAL 50 FTE.IN`	`EXTERNAL 50 FTE.IN`
`EXTERNAL 60 INEFFSW.IN`	`EXTERNAL 60 INEFFSW.IN`	`EXTERNAL 60 INEFFSW.IN`	`EXTERNAL 60 INEFFSW.IN`
`CONSTANT 0.00263`	`CONSTANT 0.00003`	`CONSTANT 0.00263`	`CONSTANT 0.00003`
`OPEN/CLOSE NRDV.IN`	`OPEN/CLOSE NRDV.IN`	`OPEN/CLOSE NRDV.IN`	`OPEN/CLOSE NRDV.IN`
`0`	`0.2`	`0`	`0.05`
`0 1   SP 3`	`0 1   SP 9`	`0 1   SP 15`	`0 1   SP 21`
`WELLS1`	`WELLS1`	`WELLS1`	`WELLS1`
`EXTERNAL 20 OFE.IN`	`EXTERNAL 20 OFE.IN`	`EXTERNAL 20 OFE.IN`	`EXTERNAL 20 OFE.IN`
`EXTERNAL 30 ROOT.IN`	`EXTERNAL 30 ROOT.IN`	`EXTERNAL 30 ROOT.IN`	`EXTERNAL 30 ROOT.IN`
`EXTERNAL 40 KC.IN`	`EXTERNAL 40 KC.IN`	`EXTERNAL 40 KC.IN`	`EXTERNAL 40 KC.IN`
`CONSTANT 0.00289`	`CONSTANT 0.00484`	`CONSTANT 0.00289`	`CONSTANT 0.00484`
`EXTERNAL 50 FTE.IN`	`EXTERNAL 50 FTE.IN`	`EXTERNAL 50 FTE.IN`	`EXTERNAL 50 FTE.IN`
`EXTERNAL 60 INEFFSW.IN`	`EXTERNAL 60 INEFFSW.IN`	`EXTERNAL 60 INEFFSW.IN`	`EXTERNAL 60 INEFFSW.IN`
`CONSTANT 0.00184`	`CONSTANT 0.00008`	`CONSTANT 0.00184`	`CONSTANT 0.00008`
`OPEN/CLOSE NRDV.IN`	`OPEN/CLOSE NRDV.IN`	`OPEN/CLOSE NRDV.IN`	`OPEN/CLOSE NRDV.IN`
`0.05`	`0.075`	`0.0125`	`0.01875`
`0 1   SP 4`	`0 1   SP 10`	`0 1   SP 16`	`0 1   SP 22`
`WELLS1`	`WELLS1`	`WELLS1`	`WELLS1`
`EXTERNAL 20 OFE.IN`	`EXTERNAL 20 OFE.IN`	`EXTERNAL 20 OFE.IN`	`EXTERNAL 20 OFE.IN`
`EXTERNAL 30 ROOT.IN`	`EXTERNAL 30 ROOT.IN`	`EXTERNAL 30 ROOT.IN`	`EXTERNAL 30 ROOT.IN`
`EXTERNAL 40 KC.IN`	`EXTERNAL 40 KC.IN`	`EXTERNAL 40 KC.IN`	`EXTERNAL 40 KC.IN`
`CONSTANT 0.00431`	`CONSTANT 0.00349`	`CONSTANT 0.00431`	`CONSTANT 0.00349`
`EXTERNAL 50 FTE.IN`	`EXTERNAL 50 FTE.IN`	`EXTERNAL 50 FTE.IN`	`EXTERNAL 50 FTE.IN`
`EXTERNAL 60 INEFFSW.IN`	`EXTERNAL 60 INEFFSW.IN`	`EXTERNAL 60 INEFFSW.IN`	`EXTERNAL 60 INEFFSW.IN`
`CONSTANT 0.00061`	`CONSTANT 0.00039`	`CONSTANT 0.00061`	`CONSTANT 0.00039`
`OPEN/CLOSE NRDV.IN`	`OPEN/CLOSE NRDV.IN`	`OPEN/CLOSE NRDV.IN`	`OPEN/CLOSE NRDV.IN`
`0.05`	`0.075`	`0.0125`	`0.01875`
`0 1   SP 5`	`0 1   SP 11`	`0 1   SP 17`	`0 1   SP 23`
`WELLS1`	`WELLS1`	`WELLS1`	`WELLS1`
`EXTERNAL 20 OFE.IN`	`EXTERNAL 20 OFE.IN`	`EXTERNAL 20 OFE.IN`	`EXTERNAL 20 OFE.IN`
`EXTERNAL 30 ROOT.IN`	`EXTERNAL 30 ROOT.IN`	`EXTERNAL 30 ROOT.IN`	`EXTERNAL 30 ROOT.IN`
`EXTERNAL 40 KC.IN`	`EXTERNAL 40 KC.IN`	`EXTERNAL 40 KC.IN`	`EXTERNAL 40 KC.IN`
`CONSTANT 0.00592`	`CONSTANT 0.00154`	`CONSTANT 0.00592`	`CONSTANT 0.00154`
`EXTERNAL 50 FTE.IN`	`EXTERNAL 50 FTE.IN`	`EXTERNAL 50 FTE.IN`	`EXTERNAL 50 FTE.IN`
`EXTERNAL 60 INEFFSW.IN`	`EXTERNAL 60 INEFFSW.IN`	`EXTERNAL 60 INEFFSW.IN`	`EXTERNAL 60 INEFFSW.IN`
`CONSTANT 0.00029`	`CONSTANT 0.00149`	`CONSTANT 0.00029`	`CONSTANT 0.00149`
`OPEN/CLOSE NRDV.IN`	`OPEN/CLOSE NRDV.IN`	`OPEN/CLOSE NRDV.IN`	`OPEN/CLOSE NRDV.IN`
`0.05`	`0`	`0.0125`	`0`
`0 1   SP 6`	`0 1   SP 12`	`0 1   SP 18`	`0 1   SP 24`
`WELLS1`	`WELLS1`	`WELLS1`	`WELLS1`
`EXTERNAL 20 OFE.IN`	`EXTERNAL 20 OFE.IN`	`EXTERNAL 20 OFE.IN`	`EXTERNAL 20 OFE.IN`
`EXTERNAL 30 ROOT.IN`	`EXTERNAL 30 ROOT.IN`	`EXTERNAL 30 ROOT.IN`	`EXTERNAL 30 ROOT.IN`
`EXTERNAL 40 KC.IN`	`EXTERNAL 40 KC.IN`	`EXTERNAL 40 KC.IN`	`EXTERNAL 40 KC.IN`
`CONSTANT 0.00688`	`CONSTANT 0.00096`	`CONSTANT 0.00688`	`CONSTANT 0.00096`
`EXTERNAL 50 FTE.IN`	`EXTERNAL 50 FTE.IN`	`EXTERNAL 50 FTE.IN`	`EXTERNAL 50 FTE.IN`
`EXTERNAL 60 INEFFSW.IN`	`EXTERNAL 60 INEFFSW.IN`	`EXTERNAL 60 INEFFSW.IN`	`EXTERNAL 60 INEFFSW.IN`
`CONSTANT 0.00013`	`CONSTANT 0.0021`	`CONSTANT 0.00013`	`CONSTANT 0.0021`
`OPEN/CLOSE NRDV.IN`	`OPEN/CLOSE NRDV.IN`	`OPEN/CLOSE NRDV.IN`	`OPEN/CLOSE NRDV.IN`
`0.2`	`0`	`0.05`	`0`

Farm Process (FMP) Ancillary Input Data Sets

Farm Wells – WELLS.IN

```
1  5   6    1  1  5
1  8   6    2  1  5
1  8   9    3  1  5
1  5   9    4  2  3
1  6  10    5  2  2
1  7  11    6  2  1
1 18   3    7  3  2
1 21   3    8  3  4
1 18  19    9  4  3
1 21  19   10  4  5
0  6   2  -11  5   5 1
0  8   2  -12  5   5 1
0 10   2  -13  5   5 1
0  3  19  -14  6  10 1
0  6  17  -15  6   6 1
```

Ground-Surface Elevation – GSE.IN

```
298.50 297.53 296.55 295.58 294.61 293.63 292.66 291.68 290.71 289.74 288.76 287.79 286.82 285.84 284.87 283.89 282.92 281.95 280.97 280.00
298.50 297.53 296.55 295.58 294.61 293.62 292.63 291.64 290.65 289.66 288.68 287.69 286.70 285.71 284.72 283.74 282.75 281.76 280.77 279.78
298.50 297.53 296.55 295.58 294.61 293.60 292.60 291.60 290.59 289.59 288.59 287.59 286.58 285.58 284.58 283.58 282.57 281.57 280.57 279.57
298.50 297.53 296.55 295.58 294.61 293.59 292.57 291.55 290.54 289.52 288.50 287.49 286.47 285.45 284.44 283.42 282.40 281.38 280.37 279.35
298.50 297.53 296.55 295.58 294.61 293.57 292.54 291.51 290.48 289.45 288.42 287.39 286.35 285.32 284.29 283.26 282.23 281.20 280.16 279.13
298.50 297.53 296.55 295.58 294.61 293.56 292.51 291.47 290.42 289.38 288.33 287.28 286.24 285.19 284.15 283.10 282.05 281.01 279.96 278.92
298.50 297.53 296.55 295.58 294.61 293.54 292.48 291.42 290.36 289.30 288.24 287.18 286.12 285.06 284.00 282.94 281.88 280.82 279.76 278.70
298.50 297.53 296.55 295.58 294.61 293.53 292.46 291.38 290.31 289.23 288.16 287.08 286.01 284.93 283.86 282.78 281.71 280.63 279.56 278.48
298.50 297.53 296.55 295.58 294.61 293.52 292.43 291.34 290.25 289.16 288.07 286.98 285.89 284.80 283.71 282.62 281.53 280.45 279.36 278.27
298.50 297.53 296.55 295.58 294.61 293.50 292.40 291.29 290.19 289.09 287.98 286.88 285.78 284.67 283.57 282.46 281.36 280.26 279.15 278.05
298.50 297.53 296.55 295.58 294.61 293.49 292.37 291.25 290.13 289.01 287.90 286.78 285.66 284.54 283.42 282.31 281.19 280.07 278.95 277.83
298.50 297.53 296.55 295.58 294.61 293.47 292.34 291.21 290.07 288.94 287.81 286.68 285.54 284.41 283.28 282.15 281.01 279.88 278.75 277.62
298.50 297.53 296.55 295.58 294.61 293.46 292.31 291.16 290.02 288.87 287.72 286.58 285.43 284.28 283.14 281.99 280.84 279.69 278.55 277.40
298.50 297.53 296.55 295.58 294.61 293.48 292.35 291.22 290.09 288.96 287.83 286.70 285.57 284.44 283.31 282.18 281.05 279.92 278.79 277.66
298.50 297.53 296.55 295.58 294.61 293.49 292.38 291.27 290.16 289.04 287.93 286.82 285.71 284.59 283.48 282.37 281.26 280.14 279.03 277.92
298.50 297.53 296.55 295.58 294.61 293.51 292.42 291.32 290.23 289.13 288.04 286.94 285.85 284.75 283.66 282.56 281.47 280.37 279.28 278.18
298.50 297.53 296.55 295.58 294.61 293.53 292.45 291.37 290.29 289.22 288.14 287.06 285.98 284.91 283.83 282.75 281.67 280.60 279.52 278.44
298.50 297.53 296.55 295.58 294.61 293.54 292.48 291.42 290.36 289.30 288.24 287.18 286.12 285.06 284.00 282.94 281.88 280.82 279.76 278.70
298.50 297.53 296.55 295.58 294.61 293.56 292.52 291.48 290.43 289.39 288.35 287.30 286.26 285.22 284.18 283.13 282.09 281.05 280.00 278.96
298.50 297.53 296.55 295.58 294.61 293.58 292.55 291.53 290.50 289.48 288.45 287.43 286.40 285.37 284.35 283.32 282.30 281.27 280.25 279.22
298.50 297.53 296.55 295.58 294.61 293.60 292.59 291.58 290.57 289.56 288.56 287.55 286.54 285.53 284.52 283.51 282.51 281.50 280.49 279.48
298.50 297.53 296.55 295.58 294.61 293.61 292.62 291.63 290.64 289.65 288.66 287.67 286.68 285.69 284.70 283.70 282.71 281.72 280.73 279.74
298.50 297.53 296.55 295.58 294.61 293.63 292.66 291.68 290.71 289.74 288.76 287.79 286.82 285.84 284.87 283.89 282.92 281.95 280.97 280.00
```

Farm-ID – FID.IN

```
7  7  7  7  7  7  7  7  7  7  7  7  7  7  7  7  7  7  7  7
7  5  5  5  7  7  7  7  7  7  7  7  7  7  7  7  7  7  7  7
7  5  5  5  7  7  7  7  7  7  7  7  7  6  6  6  6  7  7  7
7  5  5  5  7  7  2  2  2  7  7  7  6  6  6  6  7  7
7  5  5  5  7  7  2  2  2  7  7  7  6  6  6  6  7  7
7  7  7  7  1  1  1  2  2  2  7  7  7  6  6  6  6  7  7
7  7  7  7  1  1  1  1  1  1  7  7  7  6  6  6  6  7  7
7  7  7  7  1  1  1  1  1  1  7  7  7  6  6  6  6  8  8
7  7  7  7  1  1  1  1  1  1  7  7  7  6  6  6  6  8  8
7  7  7  7  7  7  7  7  7  7  7  7  7  7  7  7  7  8  8
7  7  7  7  7  7  7  7  7  7  7  7  7  7  7  7  7  8  8
7  7  7  7  7  7  7  7  7  7  7  7  7  7  7  7  7  8  8
7  7  7  7  7  7  7  7  7  7  7  7  7  8  8  8  8  8  8
7  7  7  7  7  7  7  7  7  7  7  7  7  7  7  7  7  8  8
7  7  7  7  7  7  7  7  7  7  7  7  7  7  7  7  7  8  8
7  7  7  7  7  7  7  7  7  7  7  7  7  7  7  7  7  8  8
7  7  7  7  3  3  3  3  3  3  4  4  4  4  4  7  7  8  8
7  7  7  7  7  3  3  3  3  3  4  4  4  4  4  7  7  8  8
7  7  7  7  7  3  3  3  3  3  4  4  4  4  4  7  7  7
7  7  7  7  7  7  7  7  7  7  7  7  7  7  7  7  7  7  7
7  7  7  7  7  7  7  7  7  7  7  7  7  7  7  7  7  7  7  7
7  7  7  7  7  7  7  7  7  7  7  7  7  7  7  7  7  7  7  7
7  7  7  7  7  7  7  7  7  7  7  7  7  7  7  7  7  7  7  7
```

Soil-ID – SID.IN

```
1  1  1  1  1  1  1  1  1  1  1  1  1  1  2  2  2  2  3  3
1  1  1  1  1  1  1  1  1  1  1  1  1  1  2  2  2  2  3  3
1  1  1  1  1  1  1  1  1  1  1  1  1  2  2  2  3  3  3  3
1  1  1  1  1  1  1  1  1  1  1  1  2  2  2  2  3  3  3  3
1  1  1  1  1  1  1  1  1  1  1  2  2  2  2  3  3  3  3  3
1  1  1  1  1  1  1  1  1  2  2  2  2  3  3  3  3  3  3  3
1  1  1  1  1  1  1  1  2  2  2  2  3  3  3  3  3  3  3  3
1  1  1  1  1  1  1  2  2  2  2  3  3  3  3  3  3  3  3  3
1  1  1  1  1  1  2  2  2  2  3  3  3  3  3  3  3  3  3  3
1  1  1  1  1  2  2  2  2  3  3  3  3  3  3  3  3  3  3  3
1  1  1  1  2  2  2  2  3  3  3  3  3  3  3  3  3  3  3  3
1  1  1  2  2  2  2  3  3  3  3  3  3  3  3  3  3  3  3  3
1  1  2  2  2  2  3  3  3  3  3  3  3  3  3  3  3  3  3  3
1  1  2  2  2  2  3  3  3  3  3  3  3  3  3  3  3  3  3  3
1  1  1  1  2  2  2  2  3  3  3  3  3  3  3  3  3  3  3  3
1  1  1  1  1  2  2  2  2  3  3  3  3  3  3  3  3  3  3  3
1  1  1  1  1  1  2  2  2  2  3  3  3  3  3  3  3  3  3  3
1  1  1  1  1  1  1  2  2  2  2  3  3  3  3  3  3  3  3  3
1  1  1  1  1  1  1  1  2  2  2  2  3  3  3  3  3  3  3  3
1  1  1  1  1  1  1  1  1  2  2  2  2  3  3  3  3  3  3  3
1  1  1  1  1  1  1  1  1  1  2  2  2  2  3  3  3  3  3  3
1  1  1  1  1  1  1  1  1  1  1  2  2  2  2  3  3  3  3  3
1  1  1  1  1  1  1  1  1  1  1  1  2  2  2  2  3  3  3  3
```

Soil-Type List – SOILLIST.IN
```
1 1.8 SILT
2 1.3 SANDYLOAM
3 1.5 SILTYCLAY
```

Crop-ID – CID.IN
```
5  5  5  5  5  5  5  5  5  5  5  5  5  5  5  5  5  5  5  5
5  1  2  3  5  5  5  5  5  5  5  5  5  5  5  5  5  5  5  5
5  1  2  3  5  5  5  5  5  5  5  5  5  5  5  5  5  5  5  5
5  1  2  3  5  5  5  2  3  3  5  5  5  5  4  4  4  4  5  5
5  1  2  3  5  5  5  2  3  3  5  5  5  5  4  4  4  4  5  5
5  5  5  5  1  1  2  2  3  3  5  5  5  5  4  4  4  4  5  5
5  5  5  5  1  1  2  2  3  3  5  5  5  5  4  4  4  4  5  5
5  5  5  5  1  1  2  2  3  3  5  5  5  5  4  4  4  4  6  6
5  5  5  5  1  1  2  2  3  3  5  5  5  5  4  4  4  4  6  6
5  5  5  5  5  5  5  5  5  5  5  5  5  5  5  5  5  5  6  6
5  5  5  5  5  5  5  5  5  5  5  5  5  5  5  5  5  5  6  6
5  5  5  5  5  5  5  5  5  5  5  5  5  5  5  5  5  5  6  6
5  5  5  5  5  5  5  5  5  5  5  5  5  5  6  6  6  6  6  6
5  5  5  5  5  5  5  5  5  5  5  5  5  5  5  5  5  5  6  6
5  5  5  5  5  5  5  5  5  5  5  5  5  5  5  5  5  5  6  6
5  5  5  5  5  5  5  5  5  5  5  5  5  5  5  5  5  5  6  6
5  5  5  5  5  1  1  1  1  2  2  2  2  3  3  3  5  5  6  6
5  5  5  5  5  1  1  1  2  2  2  2  3  3  3  5  5  6  6
5  5  5  5  5  5  1  1  1  2  2  2  2  3  3  3  5  5  5  5
5  5  5  5  5  5  5  5  5  5  5  5  5  5  5  5  5  5  5  5
5  5  5  5  5  5  5  5  5  5  5  5  5  5  5  5  5  5  5  5
5  5  5  5  5  5  5  5  5  5  5  5  5  5  5  5  5  5  5  5
5  5  5  5  5  5  5  5  5  5  5  5  5  5  5  5  5  5  5  5
```

Pressure Heads defining Stress Response Function – PSI.IN
(red value indicates positive pressure head at which uptake of riparian vegetation ceases)
```
1 -0.15   -0.3   -7.925 -80
2 -0.1125 -0.25  -7.375 -97.5
3 -0.005  -0.115 -11.5  -160
4 -0.15   -0.3   -7.5   -80
5 -0.1    -0.25  -5     -80
6  0.2    -0.2   -15    -100
```

Diversion Locations for Semi-Routed Delivieries – SRD.IN
```
1  3  5  2 13
2  3  8  2 16
3 17  5  6  5
4 20 12  6 15
5  0  0  0  0
6 13 15 10  4
7  0  0  0  0
8  0  0  0  0
```

Returnflow Locations for Semi-Routed Runoff Returnflow – SRR.IN
```
1  0  0  3  2
2  0  0  0  0
3  0  0  0  0
4  0  0  0  0
5  0  0  0  0
6  0  0  0  0
7  0  0  0  0
8  0  0  0  0
```

Matrix of On-Farm Efficiencies – OFE.IN

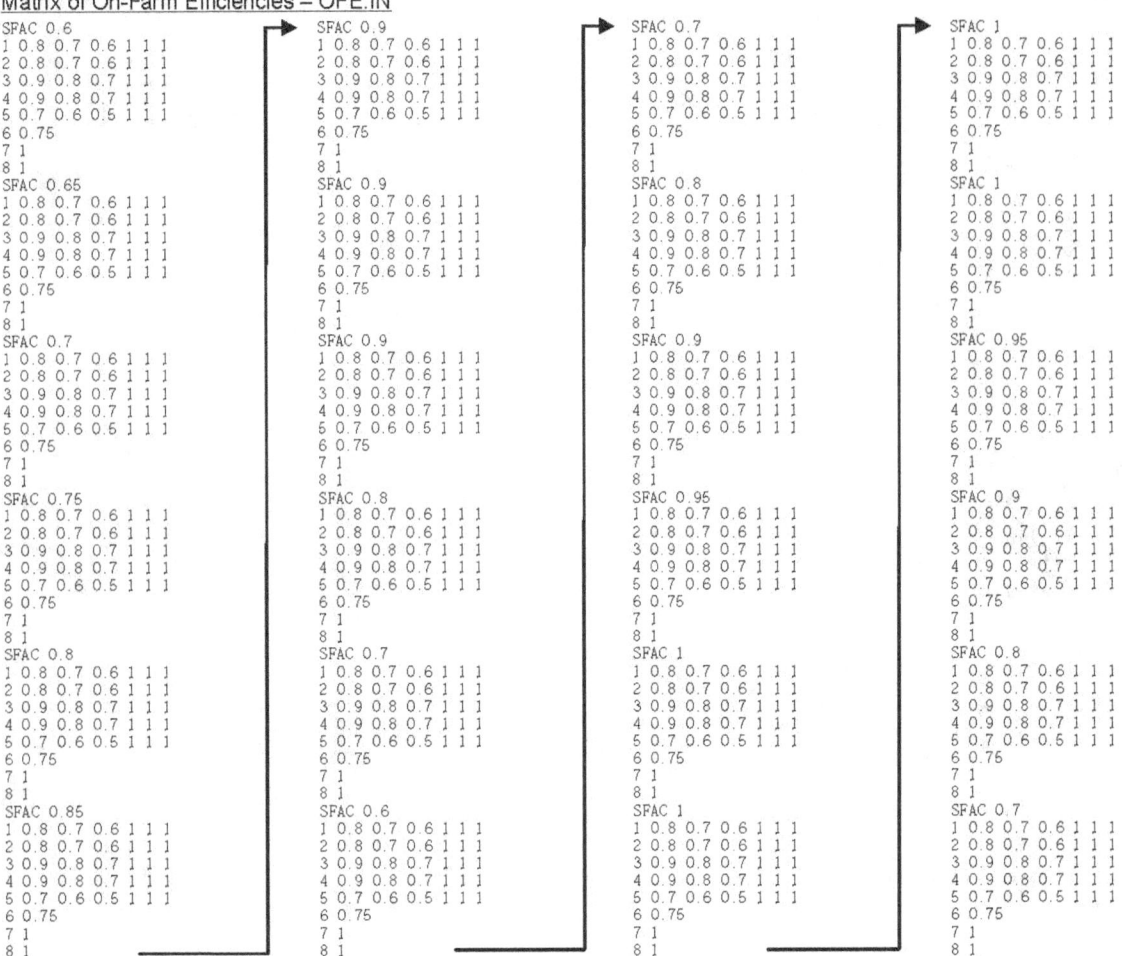

SFAC 0.6
1 0.8 0.7 0.6 1 1 1
2 0.8 0.7 0.6 1 1 1
3 0.9 0.8 0.7 1 1 1
4 0.9 0.8 0.7 1 1 1
5 0.7 0.6 0.5 1 1 1
6 0.75
7 1
8 1
SFAC 0.65
1 0.8 0.7 0.6 1 1 1
2 0.8 0.7 0.6 1 1 1
3 0.9 0.8 0.7 1 1 1
4 0.9 0.8 0.7 1 1 1
5 0.7 0.6 0.5 1 1 1
6 0.75
7 1
8 1
SFAC 0.7
1 0.8 0.7 0.6 1 1 1
2 0.8 0.7 0.6 1 1 1
3 0.9 0.8 0.7 1 1 1
4 0.9 0.8 0.7 1 1 1
5 0.7 0.6 0.5 1 1 1
6 0.75
7 1
8 1
SFAC 0.75
1 0.8 0.7 0.6 1 1 1
2 0.8 0.7 0.6 1 1 1
3 0.9 0.8 0.7 1 1 1
4 0.9 0.8 0.7 1 1 1
5 0.7 0.6 0.5 1 1 1
6 0.75
7 1
8 1
SFAC 0.8
1 0.8 0.7 0.6 1 1 1
2 0.8 0.7 0.6 1 1 1
3 0.9 0.8 0.7 1 1 1
4 0.9 0.8 0.7 1 1 1
5 0.7 0.6 0.5 1 1 1
6 0.75
7 1
8 1
SFAC 0.85
1 0.8 0.7 0.6 1 1 1
2 0.8 0.7 0.6 1 1 1
3 0.9 0.8 0.7 1 1 1
4 0.9 0.8 0.7 1 1 1
5 0.7 0.6 0.5 1 1 1
6 0.75
7 1
8 1

SFAC 0.9
1 0.8 0.7 0.6 1 1 1
2 0.8 0.7 0.6 1 1 1
3 0.9 0.8 0.7 1 1 1
4 0.9 0.8 0.7 1 1 1
5 0.7 0.6 0.5 1 1 1
6 0.75
7 1
8 1
SFAC 0.9
1 0.8 0.7 0.6 1 1 1
2 0.8 0.7 0.6 1 1 1
3 0.9 0.8 0.7 1 1 1
4 0.9 0.8 0.7 1 1 1
5 0.7 0.6 0.5 1 1 1
6 0.75
7 1
8 1
SFAC 0.9
1 0.8 0.7 0.6 1 1 1
2 0.8 0.7 0.6 1 1 1
3 0.9 0.8 0.7 1 1 1
4 0.9 0.8 0.7 1 1 1
5 0.7 0.6 0.5 1 1 1
6 0.75
7 1
8 1
SFAC 0.8
1 0.8 0.7 0.6 1 1 1
2 0.8 0.7 0.6 1 1 1
3 0.9 0.8 0.7 1 1 1
4 0.9 0.8 0.7 1 1 1
5 0.7 0.6 0.5 1 1 1
6 0.75
7 1
8 1
SFAC 0.7
1 0.8 0.7 0.6 1 1 1
2 0.8 0.7 0.6 1 1 1
3 0.9 0.8 0.7 1 1 1
4 0.9 0.8 0.7 1 1 1
5 0.7 0.6 0.5 1 1 1
6 0.75
7 1
8 1
SFAC 0.6
1 0.8 0.7 0.6 1 1 1
2 0.8 0.7 0.6 1 1 1
3 0.9 0.8 0.7 1 1 1
4 0.9 0.8 0.7 1 1 1
5 0.7 0.6 0.5 1 1 1
6 0.75
7 1
8 1

SFAC 0.7
1 0.8 0.7 0.6 1 1 1
2 0.8 0.7 0.6 1 1 1
3 0.9 0.8 0.7 1 1 1
4 0.9 0.8 0.7 1 1 1
5 0.7 0.6 0.5 1 1 1
6 0.75
7 1
8 1
SFAC 0.8
1 0.8 0.7 0.6 1 1 1
2 0.8 0.7 0.6 1 1 1
3 0.9 0.8 0.7 1 1 1
4 0.9 0.8 0.7 1 1 1
5 0.7 0.6 0.5 1 1 1
6 0.75
7 1
8 1
SFAC 0.9
1 0.8 0.7 0.6 1 1 1
2 0.8 0.7 0.6 1 1 1
3 0.9 0.8 0.7 1 1 1
4 0.9 0.8 0.7 1 1 1
5 0.7 0.6 0.5 1 1 1
6 0.75
7 1
8 1
SFAC 0.95
1 0.8 0.7 0.6 1 1 1
2 0.8 0.7 0.6 1 1 1
3 0.9 0.8 0.7 1 1 1
4 0.9 0.8 0.7 1 1 1
5 0.7 0.6 0.5 1 1 1
6 0.75
7 1
8 1
SFAC 1
1 0.8 0.7 0.6 1 1 1
2 0.8 0.7 0.6 1 1 1
3 0.9 0.8 0.7 1 1 1
4 0.9 0.8 0.7 1 1 1
5 0.7 0.6 0.5 1 1 1
6 0.75
7 1
8 1
SFAC 1
1 0.8 0.7 0.6 1 1 1
2 0.8 0.7 0.6 1 1 1
3 0.9 0.8 0.7 1 1 1
4 0.9 0.8 0.7 1 1 1
5 0.7 0.6 0.5 1 1 1
6 0.75
7 1
8 1

SFAC 1
1 0.8 0.7 0.6 1 1 1
2 0.8 0.7 0.6 1 1 1
3 0.9 0.8 0.7 1 1 1
4 0.9 0.8 0.7 1 1 1
5 0.7 0.6 0.5 1 1 1
6 0.75
7 1
8 1
SFAC 1
1 0.8 0.7 0.6 1 1 1
2 0.8 0.7 0.6 1 1 1
3 0.9 0.8 0.7 1 1 1
4 0.9 0.8 0.7 1 1 1
5 0.7 0.6 0.5 1 1 1
6 0.75
7 1
8 1
SFAC 0.95
1 0.8 0.7 0.6 1 1 1
2 0.8 0.7 0.6 1 1 1
3 0.9 0.8 0.7 1 1 1
4 0.9 0.8 0.7 1 1 1
5 0.7 0.6 0.5 1 1 1
6 0.75
7 1
8 1
SFAC 0.9
1 0.8 0.7 0.6 1 1 1
2 0.8 0.7 0.6 1 1 1
3 0.9 0.8 0.7 1 1 1
4 0.9 0.8 0.7 1 1 1
5 0.7 0.6 0.5 1 1 1
6 0.75
7 1
8 1
SFAC 0.8
1 0.8 0.7 0.6 1 1 1
2 0.8 0.7 0.6 1 1 1
3 0.9 0.8 0.7 1 1 1
4 0.9 0.8 0.7 1 1 1
5 0.7 0.6 0.5 1 1 1
6 0.75
7 1
8 1
SFAC 0.7
1 0.8 0.7 0.6 1 1 1
2 0.8 0.7 0.6 1 1 1
3 0.9 0.8 0.7 1 1 1
4 0.9 0.8 0.7 1 1 1
5 0.7 0.6 0.5 1 1 1
6 0.75
7 1
8 1

Root-Zone Depths – ROOT.IN

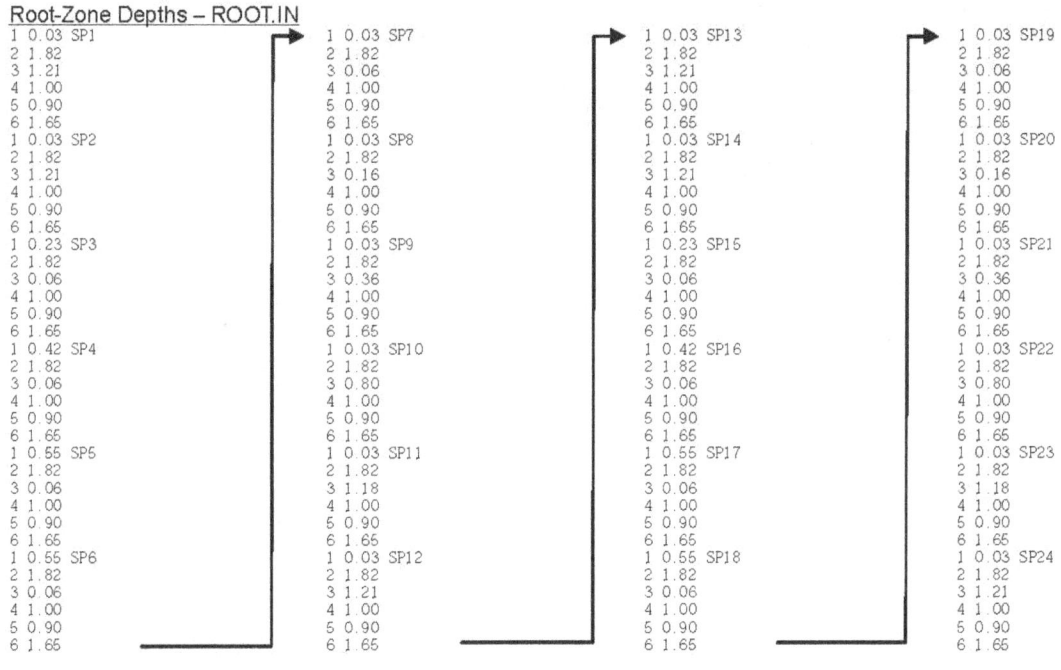

```
1 0.03 SP1        1 0.03 SP7        1 0.03 SP13       1 0.03 SP19
2 1.82            2 1.82            2 1.82            2 1.82
3 1.21            3 0.06            3 1.21            3 0.06
4 1.00            4 1.00            4 1.00            4 1.00
5 0.90            5 0.90            5 0.90            5 0.90
6 1.65            6 1.65            6 1.65            6 1.65
1 0.03 SP2        1 0.03 SP8        1 0.03 SP14       1 0.03 SP20
2 1.82            2 1.82            2 1.82            2 1.82
3 1.21            3 0.16            3 1.21            3 0.16
4 1.00            4 1.00            4 1.00            4 1.00
5 0.90            5 0.90            5 0.90            5 0.90
6 1.65            6 1.65            6 1.65            6 1.65
1 0.23 SP3        1 0.03 SP9        1 0.23 SP15       1 0.03 SP21
2 1.82            2 1.82            2 1.82            2 1.82
3 0.06            3 0.36            3 0.06            3 0.36
4 1.00            4 1.00            4 1.00            4 1.00
5 0.90            5 0.90            5 0.90            5 0.90
6 1.65            6 1.65            6 1.65            6 1.65
1 0.42 SP4        1 0.03 SP10       1 0.42 SP16       1 0.03 SP22
2 1.82            2 1.82            2 1.82            2 1.82
3 0.06            3 0.80            3 0.06            3 0.80
4 1.00            4 1.00            4 1.00            4 1.00
5 0.90            5 0.90            5 0.90            5 0.90
6 1.65            6 1.65            6 1.65            6 1.65
1 0.55 SP5        1 0.03 SP11       1 0.55 SP17       1 0.03 SP23
2 1.82            2 1.82            2 1.82            2 1.82
3 0.06            3 1.18            3 0.06            3 1.18
4 1.00            4 1.00            4 1.00            4 1.00
5 0.90            5 0.90            5 0.90            5 0.90
6 1.65            6 1.65            6 1.65            6 1.65
1 0.55 SP6        1 0.03 SP12       1 0.55 SP18       1 0.03 SP24
2 1.82            2 1.82            2 1.82            2 1.82
3 0.06            3 1.21            3 0.06            3 1.21
4 1.00            4 1.00            4 1.00            4 1.00
5 0.90            5 0.90            5 0.90            5 0.90
6 1.65            6 1.65            6 1.65            6 1.65
```

Crop Consumptive Use (here defined as Crop Coefficients) – KC.IN

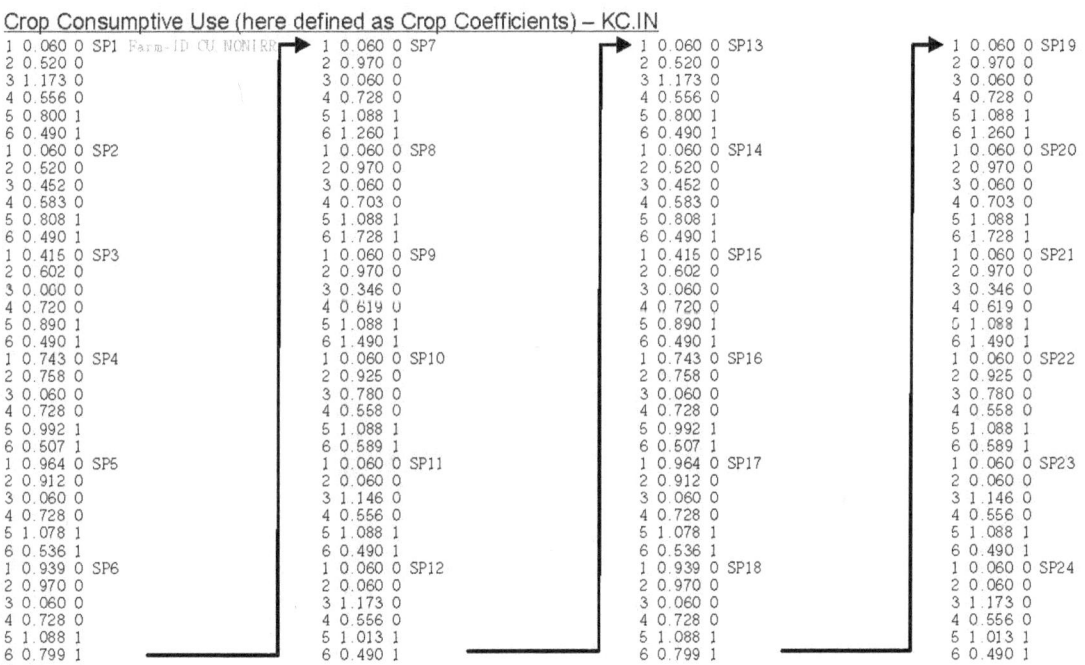

```
1 0.060 0 SP1 Farm-ID CU NONIRR   1 0.060 0 SP7      1 0.060 0 SP13     1 0.060 0 SP19
2 0.520 0                         2 0.970 0          2 0.520 0          2 0.970 0
3 1.173 0                         3 0.060 0          3 1.173 0          3 0.060 0
4 0.556 0                         4 0.728 0          4 0.556 0          4 0.728 0
5 0.800 1                         5 1.088 1          5 0.800 1          5 1.088 1
6 0.490 1                         6 1.260 1          6 0.490 1          6 1.260 1
1 0.060 0 SP2                     1 0.060 0 SP8      1 0.060 0 SP14     1 0.060 0 SP20
2 0.520 0                         2 0.970 0          2 0.520 0          2 0.970 0
3 0.452 0                         3 0.060 0          3 0.452 0          3 0.060 0
4 0.583 0                         4 0.703 0          4 0.583 0          4 0.703 0
5 0.808 1                         5 1.088 1          5 0.808 1          5 1.088 1
6 0.490 1                         6 1.728 1          6 0.490 1          6 1.728 1
1 0.415 0 SP3                     1 0.060 0 SP9      1 0.415 0 SP15     1 0.060 0 SP21
2 0.602 0                         2 0.970 0          2 0.602 0          2 0.970 0
3 0.000 0                         3 0.346 0          3 0.000 0          3 0.346 0
4 0.720 0                         4 0.619 0          4 0.720 0          4 0.619 0
5 0.890 1                         5 1.088 1          5 0.890 1          5 1.088 1
6 0.490 1                         6 1.490 1          6 0.490 1          6 1.490 1
1 0.743 0 SP4                     1 0.060 0 SP10     1 0.743 0 SP16     1 0.060 0 SP22
2 0.758 0                         2 0.925 0          2 0.758 0          2 0.925 0
3 0.060 0                         3 0.780 0          3 0.060 0          3 0.780 0
4 0.728 0                         4 0.558 0          4 0.728 0          4 0.558 0
5 0.992 1                         5 1.088 1          5 0.992 1          5 1.088 1
6 0.507 1                         6 0.589 1          6 0.507 1          6 0.589 1
1 0.964 0 SP5                     1 0.060 0 SP11     1 0.964 0 SP17     1 0.060 0 SP23
2 0.912 0                         2 0.060 0          2 0.912 0          2 0.060 0
3 0.060 0                         3 1.146 0          3 0.060 0          3 1.146 0
4 0.728 0                         4 0.556 0          4 0.728 0          4 0.556 0
5 1.078 1                         5 1.088 1          5 1.078 1          5 1.088 1
6 0.536 1                         6 0.490 1          6 0.536 1          6 0.490 1
1 0.939 0 SP6                     1 0.060 0 SP12     1 0.939 0 SP18     1 0.060 0 SP24
2 0.970 0                         2 0.060 0          2 0.970 0          2 0.060 0
3 0.060 0                         3 1.173 0          3 0.060 0          3 1.173 0
4 0.728 0                         4 0.556 0          4 0.728 0          4 0.556 0
5 1.088 1                         5 1.013 1          5 1.088 1          5 1.013 1
6 0.799 1                         6 0.490 1          6 0.799 1          6 0.490 1
```

Fractions of Transpiration and Evaporation – FTE.IN

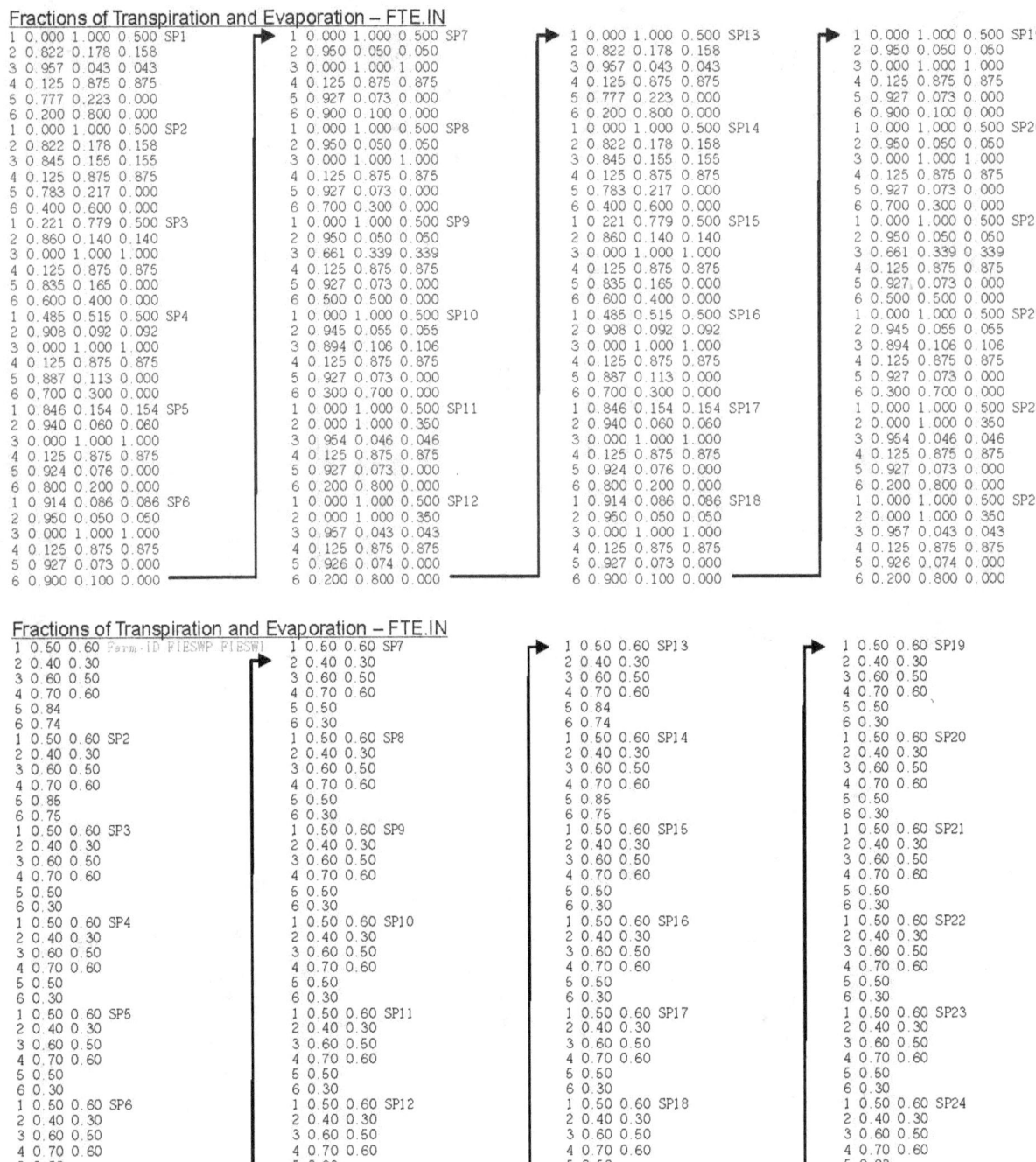

```
1 0.000 1.000 0.500 SP1      1 0.000 1.000 0.500 SP7      1 0.000 1.000 0.500 SP13     1 0.000 1.000 0.500 SP19
2 0.822 0.178 0.158          2 0.950 0.050 0.050          2 0.822 0.178 0.158          2 0.950 0.050 0.050
3 0.957 0.043 0.043          3 0.000 1.000 1.000          3 0.957 0.043 0.043          3 0.000 1.000 1.000
4 0.125 0.875 0.875          4 0.125 0.875 0.875          4 0.125 0.875 0.875          4 0.125 0.875 0.875
5 0.777 0.223 0.000          5 0.927 0.073 0.000          5 0.777 0.223 0.000          5 0.927 0.073 0.000
6 0.200 0.800 0.000          6 0.900 0.100 0.000          6 0.200 0.800 0.000          6 0.900 0.100 0.000
1 0.000 1.000 0.500 SP2      1 0.000 1.000 0.500 SP8      1 0.000 1.000 0.500 SP14     1 0.000 1.000 0.500 SP20
2 0.822 0.178 0.158          2 0.950 0.050 0.050          2 0.822 0.178 0.158          2 0.950 0.050 0.050
3 0.845 0.155 0.155          3 0.000 1.000 1.000          3 0.845 0.155 0.155          3 0.000 1.000 1.000
4 0.125 0.875 0.875          4 0.125 0.875 0.875          4 0.125 0.875 0.875          4 0.125 0.875 0.875
5 0.783 0.217 0.000          5 0.927 0.073 0.000          5 0.783 0.217 0.000          5 0.927 0.073 0.000
6 0.400 0.600 0.000          6 0.700 0.300 0.000          6 0.400 0.600 0.000          6 0.700 0.300 0.000
1 0.221 0.779 0.500 SP3      1 0.000 1.000 0.500 SP9      1 0.221 0.779 0.500 SP15     1 0.000 1.000 0.500 SP21
2 0.860 0.140 0.140          2 0.950 0.050 0.050          2 0.860 0.140 0.140          2 0.950 0.050 0.050
3 0.000 1.000 1.000          3 0.661 0.339 0.339          3 0.000 1.000 1.000          3 0.661 0.339 0.339
4 0.125 0.875 0.875          4 0.125 0.875 0.875          4 0.125 0.875 0.875          4 0.125 0.875 0.875
5 0.835 0.165 0.000          5 0.927 0.073 0.000          5 0.835 0.165 0.000          5 0.927 0.073 0.000
6 0.600 0.400 0.000          6 0.500 0.500 0.000          6 0.600 0.400 0.000          6 0.500 0.500 0.000
1 0.485 0.515 0.500 SP4      1 0.000 1.000 0.500 SP10     1 0.485 0.515 0.500 SP16     1 0.000 1.000 0.500 SP22
2 0.908 0.092 0.092          2 0.945 0.055 0.055          2 0.908 0.092 0.092          2 0.945 0.055 0.055
3 0.000 1.000 1.000          3 0.894 0.106 0.106          3 0.000 1.000 1.000          3 0.894 0.106 0.106
4 0.125 0.875 0.875          4 0.125 0.875 0.875          4 0.125 0.875 0.875          4 0.125 0.875 0.875
5 0.887 0.113 0.000          5 0.927 0.073 0.000          5 0.887 0.113 0.000          5 0.927 0.073 0.000
6 0.700 0.300 0.000          6 0.300 0.700 0.000          6 0.700 0.300 0.000          6 0.300 0.700 0.000
1 0.846 0.154 0.154 SP5      1 0.000 1.000 0.500 SP11     1 0.846 0.154 0.154 SP17     1 0.000 1.000 0.500 SP23
2 0.940 0.060 0.060          2 0.000 1.000 0.350          2 0.940 0.060 0.060          2 0.000 1.000 0.350
3 0.000 1.000 1.000          3 0.954 0.046 0.046          3 0.000 1.000 1.000          3 0.954 0.046 0.046
4 0.125 0.875 0.875          4 0.125 0.875 0.875          4 0.125 0.875 0.875          4 0.125 0.875 0.875
5 0.924 0.076 0.000          5 0.927 0.073 0.000          5 0.924 0.076 0.000          5 0.927 0.073 0.000
6 0.800 0.200 0.000          6 0.200 0.800 0.000          6 0.800 0.200 0.000          6 0.200 0.800 0.000
1 0.914 0.086 0.086 SP6      1 0.000 1.000 0.500 SP12     1 0.914 0.086 0.086 SP18     1 0.000 1.000 0.500 SP24
2 0.950 0.050 0.050          2 0.000 1.000 0.350          2 0.950 0.050 0.050          2 0.000 1.000 0.350
3 0.000 1.000 1.000          3 0.957 0.043 0.043          3 0.000 1.000 1.000          3 0.957 0.043 0.043
4 0.125 0.875 0.875          4 0.125 0.875 0.875          4 0.125 0.875 0.875          4 0.125 0.875 0.875
5 0.927 0.073 0.000          5 0.926 0.074 0.000          5 0.927 0.073 0.000          5 0.926 0.074 0.000
6 0.900 0.100 0.000          6 0.200 0.800 0.000          6 0.900 0.100 0.000          6 0.200 0.800 0.000
```

Fractions of Transpiration and Evaporation – FTE.IN

```
1 0.50 0.60 Farm-ID FIESWP FIESWI   1 0.50 0.60 SP7      1 0.50 0.60 SP13      1 0.50 0.60 SP19
2 0.40 0.30                         2 0.40 0.30          2 0.40 0.30           2 0.40 0.30
3 0.60 0.50                         3 0.60 0.50          3 0.60 0.50           3 0.60 0.50
4 0.70 0.60                         4 0.70 0.60          4 0.70 0.60           4 0.70 0.60
5 0.84                              5 0.50               5 0.84                5 0.50
6 0.74                              6 0.30               6 0.74                6 0.30
1 0.50 0.60 SP2                     1 0.50 0.60 SP8      1 0.50 0.60 SP14      1 0.50 0.60 SP20
2 0.40 0.30                         2 0.40 0.30          2 0.40 0.30           2 0.40 0.30
3 0.60 0.50                         3 0.60 0.50          3 0.60 0.50           3 0.60 0.50
4 0.70 0.60                         4 0.70 0.60          4 0.70 0.60           4 0.70 0.60
5 0.85                              5 0.50               5 0.85                5 0.50
6 0.75                              6 0.30               6 0.75                6 0.30
1 0.50 0.60 SP3                     1 0.50 0.60 SP9      1 0.50 0.60 SP15      1 0.50 0.60 SP21
2 0.40 0.30                         2 0.40 0.30          2 0.40 0.30           2 0.40 0.30
3 0.60 0.50                         3 0.60 0.50          3 0.60 0.50           3 0.60 0.50
4 0.70 0.60                         4 0.70 0.60          4 0.70 0.60           4 0.70 0.60
5 0.50                              5 0.50               5 0.50                5 0.50
6 0.30                              6 0.30               6 0.30                6 0.30
1 0.50 0.60 SP4                     1 0.50 0.60 SP10     1 0.50 0.60 SP16      1 0.50 0.60 SP22
2 0.40 0.30                         2 0.40 0.30          2 0.40 0.30           2 0.40 0.30
3 0.60 0.50                         3 0.60 0.50          3 0.60 0.50           3 0.60 0.50
4 0.70 0.60                         4 0.70 0.60          4 0.70 0.60           4 0.70 0.60
5 0.50                              5 0.50               5 0.50                5 0.50
6 0.30                              6 0.30               6 0.30                6 0.30
1 0.50 0.60 SP5                     1 0.50 0.60 SP11     1 0.50 0.60 SP17      1 0.50 0.60 SP23
2 0.40 0.30                         2 0.40 0.30          2 0.40 0.30           2 0.40 0.30
3 0.60 0.50                         3 0.60 0.50          3 0.60 0.50           3 0.60 0.50
4 0.70 0.60                         4 0.70 0.60          4 0.70 0.60           4 0.70 0.60
5 0.50                              5 0.50               5 0.50                5 0.50
6 0.30                              6 0.30               6 0.30                6 0.30
1 0.50 0.60 SP6                     1 0.50 0.60 SP12     1 0.50 0.60 SP18      1 0.50 0.60 SP24
2 0.40 0.30                         2 0.40 0.30          2 0.40 0.30           2 0.40 0.30
3 0.60 0.50                         3 0.60 0.50          3 0.60 0.50           3 0.60 0.50
4 0.70 0.60                         4 0.70 0.60          4 0.70 0.60           4 0.70 0.60
5 0.50                              5 0.83               5 0.50                5 0.83
6 0.30                              6 0.73               6 0.30                6 0.73
```

Non-Routed Surface-Water Deliveries – NRDV.IN

```
SFAC EXTERNAL 80  NRDFAC.IN
1 15000 1 0         Farm-ID NRDV NRDR NRDU
2 15000 1 0
3 20000 1 0
4 10000 1 0
5 10000 1 0
6 20000 1 0
7     0 1 0
8     0 1 0
```

Scaling for Non-Routed Surface-Water Deliveries for parameter NRDV (2nd parameter) – NRDV.IN

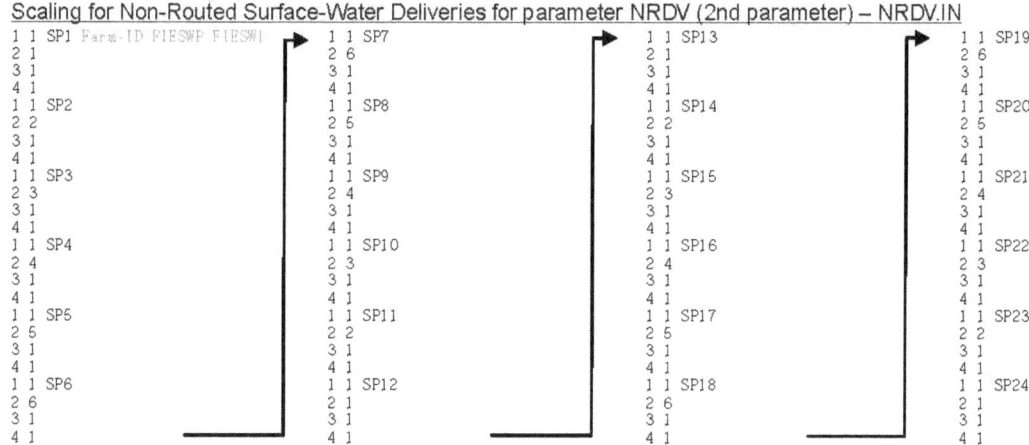

```
1 1 SP1 Farm-ID FIESWP FIESW1          1 1 SP7           1 1 SP13          1 1 SP19
2 1                                    2 6               2 1               2 6
3 1                                    3 1               3 1               3 1
4 1                                    4 1               4 1               4 1
1 1 SP2                                1 1 SP8           1 1 SP14          1 1 SP20
2 2                                    2 5               2 2               2 5
3 1                                    3 1               3 1               3 1
4 1                                    4 1               4 1               4 1
1 1 SP3                                1 1 SP9           1 1 SP15          1 1 SP21
2 3                                    2 4               2 3               2 4
3 1                                    3 1               3 1               3 1
4 1                                    4 1               4 1               4 1
1 1 SP4                                1 1 SP10          1 1 SP16          1 1 SP22
2 4                                    2 3               2 4               2 3
3 1                                    3 1               3 1               3 1
4 1                                    4 1               4 1               4 1
1 1 SP5                                1 1 SP11          1 1 SP17          1 1 SP23
2 5                                    2 2               2 5               2 2
3 1                                    3 1               3 1               3 1
4 1                                    4 1               4 1               4 1
1 1 SP6                                1 1 SP12          1 1 SP18          1 1 SP24
2 6                                    2 1               2 6               2 1
3 1                                    3 1               3 1               3 1
4 1                                    4 1               4 1               4 1
```

Hydmod (HYD) Input Data Sets

```
        22      71      -999.
BAS  HD   I    1    3750      8750 insideF2_L1
BAS  HD   I    2    3750      8750 insideF2_L2
BAS  HD   I    1    6750      4250 northofF4_L1
BAS  HD   I    2    6750      4250 northofF4_L2
SFR  SI   C    1    2      13 canal_div_f1
SFR  SO   C    1    2      13 canal_div_f1
SFR  SI   C    1    2      16 canal_div_f2
SFR  SO   C    1    2      16 canal_div_f2
SFR  SI   C    1    6       5 canal_div_f3
SFR  SO   C    1    6       5 canal_div_f3
SFR  SI   C    1    6      15 canal_div_f4
SFR  SO   C    1    6      15 canal_div_f4
SFR  SI   C    1   10       4 canal_div_f6
SFR  SO   C    1   10       4 canal_div_f6
SFR  SI   C    1    3       2 return_fl_f1
SFR  SO   C    1    3       2 return_fl_f1
SFR  SI   C    1    2      23 return_fl_f2
SFR  SO   C    1    2      23 return_fl_f2
SFR  SI   C    1    7       5 return_fl_f3
SFR  SO   C    1    7       5 return_fl_f3
SFR  SI   C    1    6      23 return_fl_f4
SFR  SO   C    1    6      23 return_fl_f4
SFR  SI   C    1   11       2 return_fl_f6
SFR  SO   C    1   11       2 return_fl_f6
```

Zonebudget (ZON) Input Data Sets

Zonebudget Batch File – Zone_Budget.bat
```
zonbud3 < FMPzone.in  > FMPzone.out
```

Zonbudget Ancillary Batch File – FMPzone.in
```
FMP_waterbalance.out csv2
cbc.out
Water Balance Regions -- 8 Farms
FMPzone.zon
A
```

Zonbudget Input file – FMPzone.zon
```
4 23 20
EXTERNAL          ()       -10
FID1.IN
EXTERNAL          ()       -10
FID2.IN
EXTERNAL          ()       -10
FID3.IN
EXTERNAL          ()       -10
FID4.IN
1  2  3  4  5  6  7  8
```

Farm Process (FMP) Routing Information Output File – Rout.out

```
ROUTING INFORMATION FOR FARM:        1, PERIOD    1
----------------------------------------------------

  DELIVERIES:
   FULLY-ROUTED DELIVERIES:
    ROUTED DELIVERY OPTION WAS NOT SELECTED
    NO ACTIVE FARM DELIVERY-SEGMENT REACHES ARE WITHIN THE FARM: NO FULLY-ROUTED DIVERSION POSSIBLE.

   SEMI-ROUTED DELIVERIES:
    SEMI-ROUTED DELIVERY FROM A SPECIFIED STREAM REACH AT:
       ROW   COLUMN   SEGMENT NO.   REACH NO.
        3      5         2            13

  RETURNFLOWS:
   FULLY-ROUTED RETURNFLOWS:
    DEACTIVATED SEARCH FOR REACHES OF ANY STREAM SEGMENTS THAT ARE WITHIN A FARM.
    NO ACTIVE FARM RETURNFLOW-SEGMENT REACHES ARE WITHIN THE FARM: NO FULLY-ROUTED RETURNFLOW POSSIBLE.

   SEMI-ROUTED RETURNFLOWS:
    SEMI-ROUTED RUNOFF-RETURNFLOW TO A SPECIFED STREAM REACH AT:
       ROW   COLUMN   SEGMENT NO.   REACH NO.
       10      8         3            2

ROUTING INFORMATION FOR FARM:        2, PERIOD    1
----------------------------------------------------

  DELIVERIES:
   FULLY-ROUTED DELIVERIES:
    ROUTED DELIVERY OPTION WAS NOT SELECTED
    NO ACTIVE FARM DELIVERY-SEGMENT REACHES ARE WITHIN THE FARM: NO FULLY-ROUTED DIVERSION POSSIBLE.

   SEMI-ROUTED DELIVERIES:
    SEMI-ROUTED DELIVERY FROM A SPECIFIED STREAM REACH AT:
       ROW   COLUMN   SEGMENT NO.   REACH NO.
        3      8         2            16

  RETURNFLOWS:
   FULLY-ROUTED RETURNFLOWS:
    ACTIVATED SEARCH FOR REACHES OF ANY STREAM SEGMENTS THAT ARE WITHIN A FARM.
    NO ACTIVE FARM RETURNFLOW-SEGMENT REACHES ARE WITHIN THE FARM: NO FULLY-ROUTED RETURNFLOW POSSIBLE.

   SEMI-ROUTED RETURNFLOWS:
    SEMI-ROUTED RUNOFF-RETURNFLOW TO A STREAM REACH FOUND NEAREST TO THE LOWEST FARM ELEVATION AT:
       ROW   COLUMN   SEGMENT NO.   REACH NO.
        6     12         2            23

ROUTING INFORMATION FOR FARM:        3, PERIOD    1
----------------------------------------------------

  DELIVERIES:
   FULLY-ROUTED DELIVERIES:
    ROUTED DELIVERY OPTION WAS NOT SELECTED
    NO ACTIVE FARM DELIVERY-SEGMENT REACHES ARE WITHIN THE FARM: NO FULLY-ROUTED DIVERSION POSSIBLE.

   SEMI-ROUTED DELIVERIES:
    SEMI-ROUTED DELIVERY FROM A SPECIFIED STREAM REACH AT:
       ROW   COLUMN   SEGMENT NO.   REACH NO.
       17      5         6            5

  RETURNFLOWS:
   FULLY-ROUTED RETURNFLOWS:
    ACTIVATED SEARCH FOR REACHES OF ANY STREAM SEGMENTS THAT ARE WITHIN A FARM.
    NO ACTIVE FARM RETURNFLOW-SEGMENT REACHES ARE WITHIN THE FARM: NO FULLY-ROUTED RETURNFLOW POSSIBLE.

   SEMI-ROUTED RETURNFLOWS:
    SEMI-ROUTED RUNOFF-RETURNFLOW TO A STREAM REACH FOUND NEAREST TO THE LOWEST FARM ELEVATION AT:
       ROW   COLUMN   SEGMENT NO.   REACH NO.
       16     11         7            5

ROUTING INFORMATION FOR FARM:        4, PERIOD    1
----------------------------------------------------

  DELIVERIES:
   FULLY-ROUTED DELIVERIES:
    ROUTED DELIVERY OPTION WAS NOT SELECTED
    NO ACTIVE FARM DELIVERY-SEGMENT REACHES ARE WITHIN THE FARM: NO FULLY-ROUTED DIVERSION POSSIBLE.

   SEMI-ROUTED DELIVERIES:
    SEMI-ROUTED DELIVERY FROM A SPECIFIED STREAM REACH AT:
       ROW   COLUMN   SEGMENT NO.   REACH NO.
       20     12         6            15

  RETURNFLOWS:
   FULLY-ROUTED RETURNFLOWS:
    ACTIVATED SEARCH FOR REACHES OF ANY STREAM SEGMENTS THAT ARE WITHIN A FARM.
    NO ACTIVE FARM RETURNFLOW-SEGMENT REACHES ARE WITHIN THE FARM: NO FULLY-ROUTED RETURNFLOW POSSIBLE.

   SEMI-ROUTED RETURNFLOWS:
    SEMI-ROUTED RUNOFF-RETURNFLOW TO A STREAM REACH FOUND NEAREST TO THE LOWEST FARM ELEVATION AT:
       ROW   COLUMN   SEGMENT NO.   REACH NO.
       17     17         6            23
```

```
ROUTING INFORMATION FOR FARM:      5, PERIOD    1
-----------------------------------------------------

  DELIVERIES:
   FULLY-ROUTED DELIVERIES:
    ROUTED DELIVERY OPTION WAS NOT SELECTED
    NO ACTIVE FARM DELIVERY-SEGMENT REACHES ARE WITHIN THE FARM: NO FULLY-ROUTED DIVERSION POSSIBLE.

   SEMI-ROUTED DELIVERIES:
    NO POINT OF DIVERSION FOR SEMI-ROUTED DELIVERY SPECIFIED: NO SEMI-ROUTED DIVERSION POSSIBLE.

  RETURNFLOWS:
   FULLY-ROUTED RETURNFLOWS:
    ACTIVATED SEARCH FOR REACHES OF ANY STREAM SEGMENTS THAT ARE WITHIN A FARM.
    FULLY ROUTED RUNOFF-RETURNFLOW PRORATED OVER REACHES WITHIN THE FARM AT:
       ROW   COLUMN   SEGMENT_NO.   REACH_NO.   REACH_LENGTH
         5       3          2           9         500.00000
         4       3          2          10         500.00000
         3       3          2          11         500.00000
         3       4          2          12         500.00000
       ACTIVE FARM RETURNFLOW-SEGMENT LENGTH:    2000.0000

   SEMI-ROUTED RETURNFLOWS:
    NO POINT OF RECHARGE FOR SEMI-ROUTED RETURNFLOW SPECIFIED: NO SEMI-ROUTED RETURNFLOW POSSIBLE.

ROUTING INFORMATION FOR FARM:      6, PERIOD    1
-----------------------------------------------------

  DELIVERIES:
   FULLY-ROUTED DELIVERIES:
    ROUTED DELIVERY OPTION WAS NOT SELECTED
    NO ACTIVE FARM DELIVERY-SEGMENT REACHES ARE WITHIN THE FARM: NO FULLY-ROUTED DIVERSION POSSIBLE.

   SEMI-ROUTED DELIVERIES:
    SEMI-ROUTED DELIVERY FROM A SPECIFIED STREAM REACH AT:
       ROW   COLUMN   SEGMENT NO.   REACH NO.
        13      15         10           4

  RETURNFLOWS:
   FULLY-ROUTED RETURNFLOWS:
    ACTIVATED SEARCH FOR REACHES OF ANY STREAM SEGMENTS THAT ARE WITHIN A FARM.
    NO ACTIVE FARM RETURNFLOW-SEGMENT REACHES ARE WITHIN THE FARM: NO FULLY-ROUTED RETURNFLOW POSSIBLE.

   SEMI-ROUTED RETURNFLOWS:
    SEMI-ROUTED RUNOFF-RETURNFLOW TO A STREAM REACH FOUND NEAREST TO THE LOWEST FARM ELEVATION AT:
       ROW   COLUMN   SEGMENT NO.   REACH NO.
        13      18         11           2

ROUTING INFORMATION FOR FARM:      7, PERIOD    1
-----------------------------------------------------

  DELIVERIES:
   FULLY-ROUTED DELIVERIES:
    ROUTED DELIVERY OPTION WAS NOT SELECTED
    NO ACTIVE FARM DELIVERY-SEGMENT REACHES ARE WITHIN THE FARM: NO FULLY-ROUTED DIVERSION POSSIBLE.

   SEMI-ROUTED DELIVERIES:
    NO POINT OF DIVERSION FOR SEMI-ROUTED DELIVERY SPECIFIED: NO SEMI-ROUTED DIVERSION POSSIBLE.

  RETURNFLOWS:
   FULLY-ROUTED RETURNFLOWS:
    ACTIVATED SEARCH FOR REACHES OF ANY STREAM SEGMENTS THAT ARE WITHIN A FARM.
    FULLY ROUTED RUNOFF-RETURNFLOW PRORATED OVER REACHES WITHIN THE FARM AT:
       ROW   COLUMN   SEGMENT_NO.   REACH_NO.   REACH_LENGTH
        13       1          1           1         500.00000
        13       2          1           2         500.00000
        13       3          1           3         250.00000
        13       3          2           1         250.00000
        12       3          2           2         500.00000
        11       3          2           3         500.00000
        10       3          2           4         500.00000
         9       3          2           5         500.00000
         8       3          2           6         500.00000
         7       3          2           7         500.00000
         6       3          2           8         500.00000
         3       5          2          13         500.00000
         3       6          2          14         500.00000
         3       7          2          15         500.00000
         3       8          2          16         500.00000
         3       9          2          17         500.00000
         3      10          2          18         500.00000
         3      11          2          19         500.00000
         3      12          2          20         500.00000
         4      12          2          21         500.00000
         5      12          2          22         500.00000
         6      12          2          23         500.00000
         7      12          2          24         500.00000
         8      12          2          25         500.00000
         9      12          2          26         500.00000
        10      12          2          27         250.00000
        10       7          3           1         250.00000
        10       8          3           2         500.00000
        10       9          3           3         500.00000
        10      10          3           4         500.00000
```

```
        10      11       3       5     500.00000
        10      12       3       6     250.00000
        10      12       4       1     250.00000
        11      12       4       2     500.00000
        12      12       4       3     500.00000
        13      12       4       4     250.00000
        13       3       5       1     250.00000
        13       4       5       2     500.00000
        13       5       5       3     250.00000
        13       5       6       1     250.00000
        14       5       6       2     500.00000
        15       5       6       3     500.00000
        16       5       6       4     500.00000
        17       5       6       5     500.00000
        18       5       6       6     500.00000
        19       5       6       7     500.00000
        20       5       6       8     500.00000
        20       6       6       9     500.00000
        20       7       6      10     500.00000
        20       8       6      11     500.00000
        20       9       6      12     500.00000
        20      10       6      13     500.00000
        20      11       6      14     500.00000
        20      12       6      15     500.00000
        20      13       6      16     500.00000
        20      14       6      17     500.00000
        20      15       6      18     500.00000
        20      16       6      19     500.00000
        20      17       6      20     500.00000
        19      17       6      21     500.00000
        18      17       6      22     500.00000
        17      17       6      23     500.00000
        16      17       6      24     250.00000
        16       7       7       1     250.00000
        16       8       7       2     500.00000
        16       9       7       3     500.00000
        16      10       7       4     500.00000
        16      11       7       5     500.00000
        16      12       7       6     500.00000
        16      13       7       7     500.00000
        16      14       7       8     500.00000
        16      15       7       9     500.00000
        16      16       7      10     500.00000
        16      17       7      11     250.00000
        16      17       8       1     250.00000
        15      17       8       2     500.00000
        14      17       8       3     500.00000
        13       5       9       1     250.00000
        13       6       9       2     500.00000
        13       7       9       3     500.00000
        13       8       9       4     500.00000
        13       9       9       5     500.00000
        13      10       9       6     500.00000
        13      11       9       7     500.00000
        13      12       9       8     250.00000
        13      12      10       1     250.00000
        13      13      10       2     500.00000
        13      14      10       3     500.00000
        13      15      10       4     500.00000
   ACTIVE FARM RETURNFLOW-SEGMENT LENGTH:    40250.000

 SEMI-ROUTED RETURNFLOWS:
  NO POINT OF RECHARGE FOR SEMI-ROUTED RETURNFLOW SPECIFIED: NO SEMI-ROUTED RETURNFLOW POSSIBLE.

ROUTING INFORMATION FOR FARM:        8, PERIOD    1
------------------------------------------------------------

DELIVERIES:
 FULLY-ROUTED DELIVERIES:
  ROUTED DELIVERY OPTION WAS NOT SELECTED
  NO ACTIVE FARM DELIVERY-SEGMENT REACHES ARE WITHIN THE FARM: NO FULLY-ROUTED DIVERSION POSSIBLE.

 SEMI-ROUTED DELIVERIES:
  NO POINT OF DIVERSION FOR SEMI-ROUTED DELIVERY SPECIFIED: NO SEMI-ROUTED DIVERSION POSSIBLE.

RETURNFLOWS:
 FULLY-ROUTED RETURNFLOWS:
  ACTIVATED SEARCH FOR REACHES OF ANY STREAM SEGMENTS THAT ARE WITHIN A FARM.
  FULLY ROUTED RUNOFF-RETURNFLOW PRORATED OVER REACHES WITHIN THE FARM AT:
    ROW   COLUMN  SEGMENT_NO.   REACH_NO.   REACH_LENGTH
     13      17         8           4       250.00000
     13      16        10           5       500.00000
     13      17        10           6       250.00000
     13      17        11           1       250.00000
     13      18        11           2       500.00000
     13      19        11           3       500.00000
     13      20        11           4       500.00000
   ACTIVE FARM RETURNFLOW-SEGMENT LENGTH:    2750.0000

 SEMI-ROUTED RETURNFLOWS:
  NO POINT OF RECHARGE FOR SEMI-ROUTED RETURNFLOW SPECIFIED: NO SEMI-ROUTED RETURNFLOW POSSIBLE.
```

Appendix B: Summary of Other Enhancements to MODFLOW-2005

By R.T. Hanson, Stanley A. Leake, and Wolfgang Schmid

Multiplier Package (MULT)

The MULT Package was modified to include exponentiation as an additional binary operator that could be performed on scalars or arrays as specified in the MULT Package input. The ability to perform exponentiation facilitates the expression of power functions for calculating vertical hydraulic conductivities. The distribution of vertical hydraulic conductivity can now uniformly grade between the harmonic and geometric mean by the specification of the power function to multiplier arrays that are based on sedimentary textural data estimated on a cell-by-cell basis (C. Brush, U.S. Geological Survey, written commun., 2006; Faunt and others, 2009c). This approach was first recognized by Belitz and others (1993) and then implemented externally in the development of the revised groundwater flow model for the central part of the western San Joaquin Valley within the Central Valley (C. Brush, U.S. Geological Survey, written commun., 2006; Faunt and others, 2009c). With this modification the estimation of power-functions of hydraulic conductivity distributions can be performed internally with MF2005. This resulted in modification of the subroutine SGWF2BAS7ARMZ within the Basic Package, GWF1BAS7.F. The exponentiation imposes absolute value on the operand; therefore the operand must be greater or equal to zero. The exponentiation operator can be positive, negative or zero. To use exponentiation in the MULT Package, the user simply uses the "^" as a binary operator similar to the other binary operators, addition, subtraction, multiplication, and division as described by Harbaugh and others (2000, p. 47-48). The MULT Package does not restrict the number of binary operations specified by the user that are performed in order from left to right. If exponentiation needs to occur prior to other mathematical operations the user should make this binary operation a separate input command prior to other mathematical operations.

Time-Series Package (HYDMOD)

The modifications to the HYDMOD Package (Hanson and Leake, 1998) allow the capture of time series from the Subsidence Package (SUB) (Hoffmann and others, 2003) and from the SFR2 Packages (Niswonger and Prudic, 2005, herein referred to as the SFR2 Package in MF2005). Input specifications of locations for retrieval of time series data in HYDMOD are the same for SUB as originally specified for the Interbed Storage Package and the same for SFR2 as the first Streamflow Routing Package (STR1). However, the input item that specifies the Package that time-series data are retrieved from (PCKG) is specified as 'SUB' for the Subsidence Package and as 'SFR' for the Streamflow Routing Package.

Streamflow Routing Package (SFR2)

The FMP1 release included a modified SFR1 Package (version 1.4) (Prudic and others, 2004) with an additional option to compute streambed elevation for reaches of SFR diversion segments, which allows the streambed slope to follow the slope of ground surface at a defined depth. This option was invoked by setting the IRDFLG flag = 1 in the SFR1 input file. Details of these changes were already documented in Schmid and others (2006). These changes to SFR1 are independent of the linkage between FMP and SFR and have been consistently applied to an accordingly modified version the Streamflow Routing Package for MODFLOW-2005, gwf1sfr7.f (SFR2, Niswonger and Prudic 2006), which is released jointly with MF2005-FMP2.

The only change made by the authors of the SFR2 Package is that the IRDFLG flag now has to be set to 2 to invoke the above described option of smoothing the streambed elevation to the ground-surface elevation at a defined depth:

IRDFLG   A flag of the SFR2, which normally is used in SFR to specify, whether input data are printed for a particular stress period are printed to the list file or not. IRDFLG can also be used to define the method of calculation for the elevation of the midpoint of a diversion segment reach. The choice of setting IRDFLG = 2 is an addendum to the GWF1SFR7 code and, therefore, is only described here and not in the SFR2 input instructions (Niswonger and Prudic, 2006, p.27). The user is referred to the SFR2 input instructions regarding the location of the IRDFLG flag, which is part of the SFR2 data input block that is read for each stress period.

0 = input data for this stress period will be printed. Elevation of top of streambed of diversion segments (canals/laterals) is interpolated between elevation of upstream and downstream ends of segments, as specified in SFR input file (SFR2 input data for a stress period are printed to the list file).

1 = input data for this stress period will not be printed. Elevation of top of streambed of diversion segments (canals/laterals) is interpolated between elevation of upstream and downstream ends of segments, as specified in SFR input file (SFR2 input data for a stress period are printed to the list file).

2 = input data for this stress period will not be printed. Streamflow Routed through a Conveyance Network to a Farm (only if SFR is specified in Name File). Elevation of top of streambed of all segments that are not diversion segments is interpolated between the elevation of the upstream and downstream ends of the segments, as specified in SFR input file. Elevation of top of streambed of diversion segments (canals/laterals) follows the slope of ground surface at a depth defined by the interpolation of:

(1) the difference between the ground surface elevation and the elevation of the upstream end of a diversion segment, as specified in the SFR input file; and

(2) the difference between the ground surface elevation and the elevation of the downstream end of a diversion segment, as specified in the SFR input file. Note limitation: IRDFLG = 2 cannot be chosen if the number of the diversion segment is equal to the total number of segments.

Unsaturated Zone Flow Package (UZF1)

UZF1 offers the option to write unformatted cell-by-cell rates of actual infiltration, ground-water recharge, evapotranspiration, and groundwater discharge to land surface to a unit number specified in IUZFCB2 (for IUZFCB2>0, and when "SAVE BUDGET" is specified in Output Control).

For the consistent use of MODFLOW's ZONEBUDGET (Harbaugh, 1990) of budget terms into and out of groundwater zones, in the MF2005-FMP2 release, the UZF1-generated actual-infiltration term was prevented from being written to cell-by-cell rates to the unit number specified in IUZFCB2 when "COMPACT BUDGET" is specified in the Name File.

Subsidence Package (SUB)

SUB offers the option to write formatted and unformatted total compaction and subsidence for non-delay and delayed beds (Hoffmann and others, 2003). These output options were expanded to facilitate the option to provide separate output of elastic and inelastic compaction and subsidence for no-delay within model layers and delayed compation within interbed systems. In addition, separate input for initial compaction for elastic and inelastic components has been included for each model layer. The changes to the input include replacing the single array of intial compaction for each model layer with an elastic and inelastic compaction specification for each model layer. The additional input variables and instructions for the initial compaction are summarized in the following revised input instructions and table B1.

Input Instructions

Input for the SUB Package is read from the file that has the type "SUB" in the name file. Optional variables are shown in brackets. All single-valued variables in data items 1, 15, and 16; layer assignments for systems of interbeds in data items 2 and 3; and material properties in data item 9 are read in free format. Data items 1, 2, 3, and 15 consist of at most one record. Two-dimensional arrays in data items 4–8 and 10–14 are read with MODFLOW-2000 utility array readers U2DREL and U2DINT. For instructions on use of array readers, refer to Harbaugh and others (2000). Changes to the original SUB packafge are delenated as — *Blue text*: Flags or parameters representing modified features from SUB and Red text: Flags or parameters representing new features of SUB. The new output control flags are summarized in Table B1.

FOR EACH SIMULATION
1. ISUBCB ISUBOC NNDB NDB NMZ NN AC1 AC2 ITMIN IDSAVE IDREST
(Enter integers for variables other than AC1 and AC2, which are floating-point variables.)

2. [LN(NNDB)] if NNDB > 0
(Enter NNDB integers separated by one or more spaces or by commas.)

3. [LDN(NDB)] if NDB > 0
(Enter NDB integers separated by one or more spaces or by commas.)

4. [RNB(NCOL,NROW)] U2DREL if NDB > 0
(One array for each of the NDB systems of interbeds.)

The following four arrays are needed to describe the initial conditions and properties of each of the NNDB systems of no-delay interbeds. All of the arrays (items 5–8) for system 1 are read first; then all of the arrays for the remaining systems.

5. [HC(NCOL,NROW)] U2DREL if NNDB > 0

6. [Sfe(NCOL,NROW)] U2DREL if NNDB > 0

7. [Sfv(NCOL,NROW)] U2DREL if NNDB > 0

8. [*ComE(NCOL,NROW)*] U2DREL if NNDB > 0

9. [*ComV(NCOL,NROW)*] U2DREL if NNDB > 0

10.[DP(NMZ,3)] if NDB > 0
(Use one record for each material zone. Data item includes NMZ records, each with a value of vertical hydraulic conductivity, elastic skeletal specific storage, and inelastic skeletal specific storage.)

The following five arrays are needed to describe the initial conditions and properties of each of the NDB systems of delay interbeds. All of the arrays (items 10-14) for system 1 are read first; then all of the arrays for the remaining systems.

11. [Dstart(NCOL,NROW)] U2DREL if NDB > 0

12. [DHC(NCOL,NROW)] U2DREL if NDB > 0

13. [*DCOME(NCOL,NROW)*] U2DREL if NDB > 0

14. [*DCOMV(NCOL,NROW)*] U2DREL if NDB > 0

15. [DZ(NCOL,NROW)] U2DREL if NDB > 0

16. [NZ(NCOL,NROW)] U2DINT if NDB > 0

17. [Ifm1 Iun1 Ifm2 Iun2 Ifm3 Iun3 Ifm4 Iun4 Ifm5 Iun5 Ifm6 Iun6 Ifm7 Iun7 Ifm8 Iun8 Ifm9 Iun9 Ifm10 Iun10]
if ISUBOC > 0. (Data item 17 consists of one record with 20 integers separated by one or more spaces or by commas.)

18.[ISP1 ISP2 ITS1 ITS2 Ifl1 Ifl2 Ifl3 Ifl4 Ifl5 Ifl6 Ifl7 Ifl8 Ifl9 Ifl10 Ifl11 Ifl12 Ifl13 Ifl14 Ifl15 Ifl16 Ifl17 Ifl18 Ifl19 Ifl20 Ifl21] if ISUBOC > 0. (Data item 18 consists of ISUBOC records with xx integers separated by one or more spaces or by commas. Please see the section entitled "Package Output" for a detailed explanation of the use of data item 18.)

Explanation of Fields Used in SUB Package Input Instructions

ISUBCB   is a flag and unit number.

If ISUBCB > 0,   it is the unit number to which cell-by-cell flow terms will be written when "SAVE BUDGET" or a non-zero value for ICBCFL is specified in MODFLOW-2000 Output Control (see Harbaugh and others, 2000, p. 52–55).

If ISUBCB ≤ 0,   cell-by-cell flow terms will not be recorded.

ISUBOC   is a flag used to control output of information generated by the SUB Package

If ISUBOC > 0,   it is the number of repetitions of item 16 to be read, each repetition of which defines a set of times steps and associated flags for printing and saving subsidence, compaction, vertical displacement, preconsolidation head, and volumetric budget.

If ISUBOC ≤ 0,   volumetric budgets for systems of delay interbeds will be printed at the end of each stress period. Subsidence, compaction, vertical displacement, and preconsolidation head will not be printed or saved.

NNDB   is the number of systems of no-delay interbeds.

NDB   is the number of systems of delay interbeds.

NMZ   is the number of material zones that are needed to define the hydraulic properties of systems of delay interbeds. Each material zone is defined by a combination of vertical hydraulic conductivity, elastic specific storage, and inelastic specific storage.

NN   is the number of nodes used to discretize the half space to approximate the head distributions in systems of delay interbeds.

AC1   is an acceleration parameter. This parameter (ω_1 in equation 25) is used to predict the aquifer head at the interbed boundaries on the basis of the head change computed for the previous iteration. A value of 0.0 results in the use of the aquifer head at the previous iteration. Limited experience indicates that optimum values may range from 0.0 to 0.6.

AC2   is an acceleration parameter. This acceleration parameter is a multiplier for the head changes to compute the head at the new iteration (ω_2 in equation 27). Values are normally between 1.0 and 2.0, but the optimum probably is closer to 1.0 than to 2.0. However, as discussed following equation 27, this parameter also can be used to help convergence of the iterative solution by using values between 0 and 1.

ITMIN   is the minimum number of iterations for which one-dimensional equations will be solved for flow in interbeds when the Strongly Implicit Procedure (SIP) is used to solve the ground-water flow equations. If the current iteration level is greater than ITMIN and the SIP convergence criterion for head closure (HCLOSE) is met at a particular cell, the one-dimensional equations for that cell will not be solved. The previous solution will be used. The value of ITMIN is read but not used if a solver other than SIP is used to solve the ground-water flow equations.

IDSAVE   is a flag and a unit number.

If IDSAVE > 0,   it is the unit number on which restart records for delay interbeds will be saved at the end of the simulation. The unit number must be associated with a BINARY data file specified in the MODFLOW Name File.

If IDSAVE > 0,   restart records for delay interbeds will not be saved.

IDREST   is a flag and a unit number.

If IDREST > 0,   it is the unit number on which restart records for delay interbeds will be read at the start of the simulation. The unit number must be associated with a BINARY data file specified in the MODFLOW Name File.

If IDREST ≤ 0,   restart records for delay interbeds will not be read.

LN   is a one-dimensional array specifying the model layer assignments for each system of no-delay interbeds. The array has NNDB values.

LDN   is a one-dimensional array specifying the model layer assignments for each system of delay interbeds. The array has NDB values.

RNB   is an array specifying the factor n_{equiv} (eq. 20) at each cell for each system of delay interbeds. The array also is used to define the areal extent of each system of interbeds. For cells beyond the areal extent of the system of interbeds, enter a number less than 1.0 in the corresponding element of this array.

HC   is an array specifying the preconsolidation head or preconsolidation stress in terms of head in the aquifer for systems of no-delay interbeds. For any model cells in which specified HC is greater than the corresponding value of starting head, the value of HC will be set to that of starting head.

Sfe   is an array specifying the dimensionless elastic skeletal storage coefficient for systems of no-delay interbeds. Values may be estimated using equation 17.

Sfv   is an array specifying the dimensionless inelastic skeletal storage coefficient for systems of no-delay interbeds. Values may be estimated using equation 17.

COME   is an array specifying the starting elastic compaction in each system of no-delay interbeds. Elastic compaction values computed by the package are added to values in this array so that printed or stored values of elastic compaction and land subsidence may include previous components. Values in this array do not affect calculations of storage changes or resulting elastic compaction. For simulations in which output values are to reflect elastic compaction and subsidence since the start of the simulation, enter zero values for all elements of this array.

COMV   is an array specifying the starting inelastic compaction in each system of no-delay interbeds. Inelastic compaction values computed by the package are added to values in this array so that printed or stored values of inelastic compaction and land subsidence may include previous components. Values in this array do not affect calculations of storage changes or resulting inelastic compaction. For simulations in which output values are to reflect inelastic compaction and subsidence since the start of the simulation, enter zero values for all elements of this array.

DP   is an array containing a table of material properties for systems of delay interbeds. For each of the NMZ zones of material properties, vertical hydraulic conductivity, elastic skeletal specific storage, and inelastic skeletal specific storage are read.

Dstart   is an array specifying starting head in interbeds for systems of delay interbeds. For a particular location in a system of interbeds, the starting head is applied to every node in the string of nodes that approximates flow in half of a doubly draining interbed.

DHC   is an array specifying the starting preconsolidation head in interbeds for systems of delay interbeds. For a particular location in a system of interbeds, the starting preconsolidation head is applied to every node in the string of nodes that approximates flow in half of a doubly draining interbed. For any location at which specified starting preconsolidation head is greater than the corresponding value of the starting head, Dstart, the value of the starting preconsolidation head will be set to that of the starting head.

DCOME   is an array specifying the starting elastic compaction in each system of delay interbeds. Elastic compaction values computed by the package are added to values in this array so that printed or stored values of elastic compaction and land subsidence may include previous components. Values in this array do not affect calculations of storage changes or resulting elastic compaction. For simulations in which output values are to reflect elastic compaction and subsidence since the start of the simulation, enter zero values for all elements of this array.

DCOMV   is an array specifying the starting inelastic compaction in each system of delay interbeds. Inelastic compaction values computed by the package are added to values in this array so that printed or stored values of inelastic compaction and land subsidence may include previous components. Values in this array do not affect calculations of storage changes or resulting inelastic compaction. For simulations in which output values are to reflect inelastic compaction and subsidence since the start of the simulation, enter zero values for all elements of this array.

DZ   is an array specifying the equivalent thickness for a system of delay interbeds (b_{equiv} in equation 19).

NZ   is an array specifying the material zone numbers for systems of delay interbeds. The zone number for each location in the model grid selects the hydraulic conductivity, elastic skeletal specific storage, and inelastic skeletal specific storage of the interbeds.

Ifm1   is a code for the format in which subsidence will be printed. Format codes for variables Ifm1— Ifm9 are as follows:

0	- (10G11.4)	7	- (20F5.0)
1	- (11G10.3)	8	- (20F5.1)
2	- (9G13.6)	9	- (20F5.2)
3	- (15F7.1)	10	- (20F5.3)
4	- (15F7.2)	11	- (20F5.4)
5	- (15F7.3)	12	- (10G11.4)
6	- (15F7.4)		

Iun1   is the unit number to which subsidence will be written if it is saved on disk.

Ifm2   is a code for the format in which total compaction by model layer will be printed.

Iun2   is the unit number to which total compaction by model layer will be written if it is saved on disk.

Ifm3   is a code for the format in which total compaction by interbed system will be printed.

Iun3   is the unit number to which total compaction by interbed system will be written if it is saved on disk.

Ifm4   is a code for the format in which vertical displacement will be printed.

Iun4   is the unit number to which vertical displacement will be written if it is saved on disk.

Ifm5   is a code for the format in which no-delay preconsolidation head will be printed.

Iun5   is the unit number to which no-delay preconsolidation head will be written if it is saved on disk.

Ifm6   is a code for the format in which delay preconsolidation head will be printed.

Iun6   is the unit number to which delay preconsolidation head will be written if it is saved on disk.

Ifm7   is a code for the format in which elastic compaction by model layer will be printed.

Iun7   is the unit number to which elastic compaction by model layer will be written if it is saved on disk.

Ifm8   is a code for the format in which inelastic compaction by model layer will be printed.

Iun8   is the unit number to which inelastic compaction by model layer will be written if it is saved on disk.

Ifm9   is a code for the format in which elastic compation by interbed system will be printed.

Iun9   is the unit number to which elastic compation by interbed system will be written if it is saved on disk.

Ifm10   is a code for the format in which inelastic compation by interbed system will be printed.

Iun10   is the unit number to which inelastic compation by interbed system will be written if it is saved on disk.

The variables ISP1, ISP2, ITS1, ITS2, and Ifl1 through Ifl21 are used to control printing and saving of information generated by the SUB Package during program execution. The use of some of these variables is explained in more detail in the section entitled Package Output. The default condition for flags Ifl1 through Ifl21 is to not print or save the indicated information, except for printing budgets for no-delay interbeds for the last time step of each stress period.

ISP1   is the starting stress period in the range of stress periods to which output flags Ifl1 through Ifl21 apply. If the value of ISP1 is less than 1, the SUB Package will change the number to 1.

ISP2   is the ending stress period in the range of stress periods and time steps to which output flags Ifl1 through Ifl21 apply. If the value of ISP1 is greater than NPER (the number of stress periods in the simulation), the SUB Package will change the number to NPER.

ITS1   is the starting time step in the range of time steps in each of the stress periods ISP1 through ISP2 to which output flags Ifl1 through Ifl21 apply. If the value of ITS1is less than 1, the SUB Package will change the number to 1.

ITS2   is the ending time step in the range of time steps in each of stress periods ISP1 through ISP2 to which output flags Ifl1 through Ifl21 apply. If the value of ITS2 is greater than the number of time steps in a given stress period, the SUB Package will change the number to the number of time steps in that stress period.

Ifl1   is the output flag for printing subsidence for the set of time steps specified by ISP1, ISP2, ITS1, and ITS2.
        If Ifl1 < 0,   use default or previously defined settings of Ifl1 for printing subsidence.
        If Ifl1 = 0,   do not print subsidence.
        If Ifl1 > 0,   print subsidence.

Ifl2   is the output flag for saving subsidence to an unformatted disk file for the set of time steps specified by ISP1, ISP2, ITS1, and ITS2.
    If Ifl2 < 0,   use default or previously defined settings of Ifl2 for saving subsidence.
    If Ifl2 = 0,   do not save subsidence.
    If Ifl2 > 0,   save subsidence.

Ifl3   is the output flag for printing total compaction by model layer for the set of time steps specified by ISP1, ISP2, ITS1, and ITS2.
    If Ifl3 < 0,   use default or previously defined settings of Ifl3 for printing compaction by model layer.
    If Ifl3 = 0,   do not print compaction by model layer.
    If Ifl3 > 0,   print compaction by model layer.

Ifl4   is the output flag for saving total compaction by model layer to an unformatted disk file for the set of time steps specified by ISP1, ISP2, ITS1, and ITS2.
    If Ifl4 < 0,   use default or previously defined settings of Ifl4 for saving compaction by model layer.
    If Ifl4 = 0,   do not save compaction by model layer.
    If Ifl4 > 0,   save compaction by model layer.

Ifl5   is the output flag for printing total compaction by interbed system for the set of time steps specified by ISP1, ISP2, ITS1, and ITS2.
    If Ifl5 < 0,   use default or previously defined settings of Ifl5 for printing compaction by interbed system.
    If Ifl5 = 0,   do not print compaction by interbed system.
    If Ifl5 > 0,   print compaction by interbed system.

Ifl6   is the output flag for saving total compaction by interbed system to an unformatted disk file for the set of time steps specified by ISP1, ISP2, ITS1, and ITS2.
    If Ifl6 < 0,   use default or previously defined settings of Ifl6 for saving compaction by interbed system.
    If Ifl6 = 0,   do not save compaction by interbed system.
    If Ifl6 > 0,   save compaction by interbed system.

Ifl7   is the output flag for printing vertical displacement for the set of time steps specified by ISP1, ISP2, ITS1, and ITS2.
    If Ifl7 < 0,   use default or previously defined settings of Ifl7 for printing vertical displacement.
    If Ifl7 = 0,   do not print vertical displacement.
    If Ifl7 > 0,   print vertical displacement.

Ifl8   is the output flag for saving vertical displacement to an unformatted disk file for the set of time steps specified by ISP1, ISP2, ITS1, and ITS2.
    If Ifl8 < 0,   use default or previously defined settings of Ifl8 for saving vertical displacement.
    If Ifl8 = 0,   do not save vertical displacement.
    If Ifl8 > 0,   save vertical displacement.

Ifl9   is the output flag for printing critical head for no-delay interbeds for the set of time steps specified by ISP1, ISP2, ITS1, and ITS2.
    If Ifl9 < 0,   use default or previously defined settings of Ifl9 for printing critical head for no-delay interbeds.
    If Ifl9 = 0,   do not print critical head for no-delay interbeds.
    If Ifl9 > 0,   print critical head for no-delay interbeds.

Ifl10   is the output flag for saving critical head for no-delay interbeds to an unformatted disk file for the set of time steps specified by ISP1, ISP2, ITS1, and ITS2.
    If Ifl10 < 0,   use default or previously defined settings of Ifl10 for saving critical head for no-delay interbeds.
    If Ifl10 = 0,   do not save critical head for no-delay interbeds.
    If Ifl10 > 0,   save critical head for no-delay interbeds.

Ifl11   is the output flag for printing critical head for delay interbeds for the set of time steps specified by ISP1, ISP2, ITS1, and ITS2.
    If Ifl11 < 0,   use default or previously defined settings of Ifl11 for printing critical head for delay interbeds.
    If Ifl11 = 0,   do not print critical head for delay interbeds.
    If Ifl11 > 0,   print critical head for delay interbeds.

Ifl12   is the output flag for saving critical head for delay interbeds to an unformatted disk file for the set of time steps specified by ISP1, ISP2, ITS1, and ITS2.
     If Ifl12 < 0,   use default or previously defined settings of Ifl12 for saving critical head for delay interbeds.
     If Ifl12 = 0,   do not save critical head for delay interbeds.
     If Ifl12 > 0,   save critical head for delay interbeds.

Ifl13   is the output flag for printing volumetric budget for delay interbeds for the set of time steps specified by ISP1, ISP2, ITS1, and ITS2.
     If Ifl13 < 0,   use default or previously defined settings of Ifl13 for printing volumetric budget for delay interbeds.
     If Ifl13 = 0,   do not print volumetric budget for delay interbeds.
     If Ifl13 > 0,   print volumetric budget for delay interbeds.

Ifl14   is the output flag for printing elastic compaction by model layer for delay or no-delay interbeds for the set of time steps specified by ISP1, ISP2, ITS1, and ITS2.
     If Ifl14 < 0,   use default or previously defined settings of Ifl14 for printing elastic compaction for delay interbeds.
     If Ifl14 = 0,   do not print elastic compaction for delay interbeds.
     If Ifl14 > 0,   print elastic compaction for delay interbeds.

Ifl15   is the output flag for saving elastic compaction by model layer for delay or no-delay interbeds for the set of time steps specified by ISP1, ISP2, ITS1, and ITS2.
     If Ifl15 < 0,   use default or previously defined settings of Ifl15 for printing elastic compaction for delay interbeds.
     If Ifl15 = 0,   do not print elastic compaction for delay interbeds.
     If Ifl15 > 0,   print elastic compaction for delay interbeds.

Ifl16   is the output flag for printing inelastic compaction by model layer for delay or no-delay interbeds for the set of time steps specified by ISP1, ISP2, ITS1, and ITS2.
     If Ifl16 < 0,   use default or previously defined settings of Ifl16 for printing inelastic compaction for delay interbeds.
     If Ifl16 = 0,   do not print inelastic compaction for delay interbeds.
     If Ifl16 > 0,   print inelastic compaction for delay interbeds.

Ifl17   is the output flag for saving inelastic compation by model layer for delay or no-delay interbeds for the set of time steps specified by ISP1, ISP2, ITS1, and ITS2.
     If Ifl17 < 0,   use default or previously defined settings of Ifl17 for printing inelastic compaction for delay interbeds.
     If Ifl17 = 0,   do not print inelastic compaction for delay interbeds.
     If Ifl17 > 0,   print inelastic compaction for delay interbeds.

Ifl18   is the output flag for printing elastic compaction by interbed system for delay or no-delay interbeds for the set of time steps specified by ISP1, ISP2, ITS1, and ITS2.
     If Ifl18 < 0,   use default or previously defined settings of Ifl18 for printing elastic compaction for delay interbeds.
     If Ifl18 = 0,   do not print elastic compaction for delay interbeds.
     If Ifl18 > 0,   print elastic compaction for delay interbeds.

Ifl19   is the output flag for saving elastic compaction by interbed system for delay or no-delay interbeds for the set of time steps specified by ISP1, ISP2, ITS1, and ITS2.
     If Ifl19 < 0,   use default or previously defined settings of Ifl19 for printing elastic compaction for delay interbeds.
     If Ifl19 = 0,   do not print elastic compaction for delay interbeds.
     If Ifl19 > 0,   print elastic compaction for delay interbeds.

Ifl20   is the output flag for printing inelastic compaction by interbed system for delay or no-delay interbeds for the set of time steps specified by ISP1, ISP2, ITS1, and ITS2.
     If Ifl20 < 0,   use default or previously defined settings of Ifl20 for printing inelastic compaction for delay interbeds.
     If Ifl20 = 0,   do not print inelastic compaction for delay interbeds.
     If Ifl20 > 0,   print inelastic compaction for delay interbeds.

Ifl21   is the output flag for saving inelastic compation by interbed system for delay or no-delay interbeds for the set of time steps specified by ISP1, ISP2, ITS1, and ITS2.
     If Ifl21 < 0,   use default or previously defined settings of Ifl21 for printing inelastic compaction for delay interbeds.
     If Ifl21 = 0,   do not print inelastic compaction for delay interbeds.
     If Ifl21 > 0,   print inelastic compaction for delay interbeds.

Table B1. Information that can be optionally printed or saved by the Subsidence Package and associated variable names, numbers of arrays, and array names.

Information to be printed or saved	Variable containing print format in input data item 15	Variable containing unit number in input data item 15	Variable containing flag in data item 16 indicating print action	Variable containing flag in data item 16 indicating save action	Number of layer arrays that will be printed or saved each time step	Name of array as listed in printout and in header record of saved array
Subsidence	Ifm1	Iun1	Ifl1	Ifl2	1	SUBSIDENCE
Total compaction by model layer	Ifm2	Iun2	Ifl3	Ifl4	One array for each layer with delay or no-delay interbeds	LAYER COMPACTION
Total compaction by interbed system	Ifm3	Iun3	Ifl5	Ifl6	NNDB+NDB	NDSYS COMPACTION or DSYS COMPACTION
Vertical displacment by model layer	Ifm4	Iun4	Ifl7	Ifl8	NLAY	Z DISPLACEMENT
Critical head for no-delay interbedsr	Ifm5	Iun5	Ifl9	Ifl10	One array for each layer with no-delay interbeds	ND CRITICAL HEAD
Critical head for delay interbeds	Ifm6	Iun6	Ifl11	Ifl12	NDB	D CRITICAL HEAD
Volumetric budget for delay interbeds	—	—	Ifl13	—	—	—
Elastic compaction by model layer	Ifm7	Iun7	Ifl14	Ifl15	One array for each layer with delay or no-delay interbeds	EL LAYER CMPT
Inelastic compaction by model layer	Ifm8	Iun8	Ifl116	Ifl17	One array for each layer with delay or no-delay interbeds	INEL LAYER CMPT
Elastic compaction by interbed system	Ifm9	Iun9	Ifl18	Ifl19	NNDB+ NDB	NDSYS EL CMPT or DSYS EL CMPT
Inelastic compaction by interbed system	Ifm10	Iun10	Ifl20	Ifl21	NNDB+NDB	NDSYS INEL CMPT or DSYS INEL CMPT

www.ingramcontent.com/pod-product-compliance
Lightning Source LLC
Chambersburg PA
CBHW081503170526
45166CB00008B/2537